Food Tech Transitions

Cinzia Piatti
Simone Graeff-Hönninger • Forough Khajehei
Editors

Food Tech Transitions

Reconnecting Agri-Food, Technology
and Society

Editors
Cinzia Piatti
Department of Societal
Transition and Agriculture
University of Hohenheim
Stuttgart, Germany

Simone Graeff-Hönninger
Institute of Crop Science
University of Hohenheim
Stuttgart, Germany

Forough Khajehei
Institute of Crop Science
University of Hohenheim
Stuttgart, Germany

ISBN 978-3-030-21061-8   ISBN 978-3-030-21059-5   (eBook)
https://doi.org/10.1007/978-3-030-21059-5

This Springer imprint is published by the registered company Springer Nature Switzerland AG
The registered company address is: Gewerbestrasse 11, 6330 Cham, Switzerland

Acknowledgments

The experience of editing a book is internally challenging but rewarding. When the three of us embarked on the journey of editing and developing this book, we did not know what to expect as the three of us had never worked on a project like this together. While it was a joyful, challenging, but nevertheless stressful period, we appreciated the experience of being motivated by each other, learning from each other, working on strategies to deal with different arduous situations, getting to know each other better, and deepening our friendship to a fountain of full inspiration beyond the book.

We especially want to thank the individuals who helped make this happen. Overall, we want to thank the team of Springer, especially Sabina Ashbaugh and Sofia Valsendur, for the trust in us to develop this book and the idea into a successful story. Our sincere gratitude goes to the contributors and reviewers who contributed their time and expertise to this book. We especially thank them for staying in time and delivering the chapters as promised. That made our work and duties as editors much more pleasant.

Cinzia is very grateful to and wants to thank three people in particular: her coeditors, for being supportive and present during the different stages in this book, and Prof. Claudia Bieling, for the time and space she allowed for growing professionally, this made the difference.

Forough wants to express her deepest appreciation toward Cinzia and Simone for their invaluable support and guidance during the development of this book. She is indeed fortunate to have the chance to learn from them.

Simone is eternally grateful to Cinzia and Forough, who covered most of the work, especially their understanding for the periods she could not continuously contribute to the book. Major thanks also go to her university working group for being patient and insightful when other issues had to wait and the work on the book was a priority.

Contents

Part I   Food Tech, Raw Materials and Trends

1   Novel Food Technologies and Their Acceptance   3
Forough Khajehei, Cinzia Piatti, and Simone Graeff-Hönninger

2   Overview on the Food Industry and Its Advancement   23
Mehrdad Niakousari, Sara Hedayati, Zahra Tahsiri,
and Hamide Mirzaee

3   Post-Harvest Treatments and Related Food Quality   49
Bernhard Trierweiler and Christoph H. Weinert

**4   Nutritional and Industrial Relevance of Particular
Neotropical Pseudo-cereals**   65
Catalina Acuña-Gutiérrez, Stefanny Campos-Boza,
Andrés Hernández-Pridybailo, and Víctor M. Jiménez

**5   The Demand for Superfoods: Consumers' Desire, Production
Viability and Bio-intelligent Transition**   81
Simone Graeff-Hönninger and Forough Khajehei

Part II   Food Tech, Society and Industry

**6   Altering Production Patterns in the Food Industry:
3D Food Printing**   97
Ioannis Skartsaris and Cinzia Piatti

7   Food Consumption and Technologies   111
Cinzia Piatti and Forough Khajehei

8 Technologies at the Crossroads of Food Security and Migration.... 129
Lubana Al-Sayed

9 Corporate Responsibility in a Transitioning Food Environment: Truth-Seeking and Truth-Telling............................ 149
Louise Manning

Conclusions 171

Introduction

Prologue

This book was thought to contribute to the effort of bridging the main domains of academic knowledge in the agricultural field and provide a perspective on the advancement of food technology. The perspective provided here is not merely technical (also in relation to farming practices) and not exclusively social. The authors who contributed to this book come from agronomy, crop science, food technology, and of course social science and have come together to provide reflections on the food technology issues of our contemporary age. In our academic practice, we realized that often we have a fragmented understanding and vision of a sector or of a discipline, not to mention the narrow vision of main events and historical progression of our societies. Such fragmentation is often a consequence of our specializations, focused on a specific subject, which might impinge on a thorough understanding of agri-food issues. To make sense of the state of the art of food technology, there are many good textbooks, but they are of course sectoral books. This book, thought for our students, is at the crossroad of different fields in agri-food and aims at providing support to grasp the effects of some social advancement in agricultural production or, vice versa, how is innovation in a specific sector driving social change. Of course, which one precedes the other is a matter of specific occurrences, and we state the obvious when we repeat that the two are interrelated.

This book, while offering some insights into this, aims at inviting a discussion among professionals in agri-food about food technology in this historical moment, which some colleagues have defined as a transition age because of the many uncertainties in political and economic terms, and for the massive role that technology has come to play in both our private and professional lives. We are aware that the pace of innovation and social change will force new debates soon and will open up new paths in our academic fields.

Why Food Technology Transitions?

This book is designed to integrate knowledge about food-related technology with social sciences and a wider social perspective. The book tackles the advancement in food technologies that affects agronomic practices and also meals creation through the lens of wider social issues, analyzing the implications and challenges we are currently facing. We propose that a critical analysis of food technology cannot avoid questioning the impact it implies, with the disruption potential of agronomic practices, production and consumption logics, nutrition patterns, and human and environmental ethics that are associated to it. To explain why "food tech transitions," we have to take two separate steps, starting from agri-food, in which we enucleate food tech, and then moving to technology and transitions.

(Agri-)Food and Technology

The term "agri-food" stems from and is used to highlight the development of rural social research starting from the 1980s (see Buttel and Newby 1980; Campbell 2016), but it has been employed freely in many contexts. Basically, we can differentiate three main domains which belong to agri-food: production, processing and supply/distribution, and finally consumption. Production has been for long the focus of agricultural studies, as societies were politically and economically built on the belief that everything revolved around production (Marsden 1989). Maximizing and perfecting it, ensuring that all the issues around production were under control, would have led to prosperity, for all. Whether in a Keynasian or liberist fashion, at least since post-World War II, production has been the key for social stability. This was valid for industry and manufacturing, as well as for agriculture. It has been quite a recent discovery that consumption as well has some importance, especially in agri-food (Goodman and DuPuis 2002). To recognize that the other half of food is for consumption, which has to be taken into account, marked a process of dislocation from limited understandings to a wider gaze over power dynamics, and to recognize that, consumers are embedded in social relations (Carolan 2012:281). The existence of a broader context within which food exists must be taken into account. As a result, the relationship between production and consumption and general food relation gains is of increasing importance in the analysis. It is though slightly different when it comes to the 'middle step', that is, processing and distribution. Our attention is for this step as the middle part of this continuum that runs from production to consumption and to which food tech belongs. Food technologies in the guise of food processing date back 2 million years ago with its oldest mode, namely, cooking (Wrangham 2009). Moving through prehistory on, other forms of food processing were framed to preserve food materials in different forms of food processing such as fermentation, drying, and preservation of food resources by additives such as salt. Therefore, our ancestors first learned how to cook the food material and then how to transform, preserve, and store them. Food processing and preservation enabled the survival of groups, communities, and societies through history. At the end of the last ice age, by evolution of domestication of plants and cultivation methods, followed by domestication of animals, the plant and animal agriculture transformed and enhanced the growth of human communities and conditions. The experience-based knowledge of food processing which was gained through time succored humans to overcome hunger not only by harvesting food material provided by agriculture but also by their preservation (Floros et al. 2010). Although processing, understood as an activity related to transforming food from its raw status into a new one, can be traced back to prehistory, there is agreement that a more modern form of processing which includes fermenting, drying, and early forms of preserving was common knowledge for ancient societies in a quite ubiquitous way. All this happened mainly at the household level or in apt places such as workshops. From the invention and discoveries of Appert, Pasteur, or Liebig, to name just a few of the most famous scientists, so renowned and familiar to food tech and agri-food scholars, who developed important techniques and practices, the past centuries have seen an incredible advancement in terms of food processing (Stewart and Amerine 2012). And it is in the times of these discoveries that we have started to differentiate between household and industrial food, as it is after the Industrial Revolution that food technologies develop (Truninger 2013). Many of the changes in agri-food happened in terms of food being taken from the fields and "worked" before being distributed and made available for consumption, and the most important change for understanding modern food provision can be located historically again after World War II. We cannot stress enough the importance of this historical period, because it was the post-war compromise for promoting peace and stability that provided first the impetus for transformation and restructuring in the agricultural sector ushering in the green revolution as the main event which changed food production for good (Campbell 2012); and secondly the many changes in food consumption to the advancement and solutions of the food processing industry starting from mid-1950s (Pilcher 2016). That political configuration has now gone, and this goes together with the evolution of the production sector and the consumption politics which are quite central for the theme of this book. Indeed, it is in the aftermath of postwar that the term "industry" starts to be applied to agriculture and the food sector in a wider way (Friedmann and McMichael 1989). And in the next 20 years, the products of the food industry, especially of western and specifically American ones, become widely available, affecting meal patterns, supply, habits, and urban development. Thirty to fifty years after the end of war, we saw the then nascent food industry evolving in ways never envisaged before, with a huge focus on specific commodities which are disassembled and reassembled in new forms and whose components are "broken down into dozens of products (...) which found their way into thousands of processed foods" (Bryant et al. 2013:42–43). Further expansion of innovations based on science and technologies during the twenty-first century is nurturing the novel achievements in the thermal and non thermal food processing and technologies including volumetric heating methods (microwave, radiofrequency, ohmic heating, etc.) or non thermal processes such as high pressure processing and pulsed electric field to replace the conventional methods of processing such as pasteurization and sterilization in the food industry (Floros et al. 2010). Food technologies have been an integral part of the processing industry since the beginning but are now facing a new phase that reflects the transition age we are immersed in. In fact, this millennium has seen the development of agri-food toward provision of healthy, traditional, ethical, environmentally friendly food products: the literature is ripe with research on sustainability, food provenance, and food trends as a result of the massive demand for these food products. Production had to rise to these demands. How technology fits into this last period and matches production issues is one of the key concerns of this book. There is a tension that has to be unveiled in here because production relies on many of the commodities that come from the Global South and are intended to provide for the needs of consumers in the Global North. These relationships date back centuries, inaugurating the age of colonialism (Friedmann and McMichael 1989) and the contemporary organization of food provisioning, which have become unsustainable. The new commodities available in our markets that fulfill our desires for healthy food do not provide for amelioration of these relations, and neither of environmental conditions, for what that matters.

Technological Embeddedness

Technology has been the focus of a specific branch of studies, the science and technology studies (STS), but it has been widely researched from different perspectives. Truninger (2013:87) reminds us that STS "was the result of cross-fertilization among the sociology of scientific knowledge (SSK), the sociology of technology, and the history of technology and science," and as such, it has included the many themes of sociological, scientific, and historical inquiries. Research has addressed digitization (Peters 2016), informatization (Kallinikos 2006), datafication (van Dijck 2014), and lately also platformization (Helmond 2015). In agri-food as well, there is a long tradition of research on tech along the classical rural sociology themes of political economy and therefore production, capitalist accumulation and labor, and related themes such as ecology, genetically modified organisms, alternative food networks, participation, governance, and neoliberalism (for a lucid overview, see Legun 2016). The World Economic Forum (2017) maintains that some of the technologies now available "could be game-changing for food systems, contributing to radically new approaches along the agricultural value chain and beyond." A good amount of the contemporary agri-food debates and research on technology revolves on the so-called Farming 4.0 and focuses on data ownership, machines full connectivity, and role of farmers (see, e.g., Carolan 2017; Michalopoulos 2015; Fraser 2018; Silva et al. 2011). The evolution of application of technology and data began in the 1990s. Yield mapping and different rates of fertilizer application have been around since then, moving into precision agriculture in a massive way at the dawn of the millennium (for those interested in a revealing anticipation of these issues back to 1990s, it is worth reading Wolf and Buttel 1996 and Wolf and Wood

1997). If the first decade of the 2000s was technologically devoted to finding integrated solutions for farmers, it is only in this last decade that we have started talking about data and have the ability to combine all of them, thanks to the advent of full connectivity of machines, tablets, and last-generation smartphones, for which user interface is paramount. In fact, "in addition to its ancillary role in delivering conventional goods and services, data has intrinsic value in developing artificial intelligence (AI) capabilities and in enabling targeted marketing" (Ciuriak and Ptashkina 2018: vi), which means that we are just at the beginning of a production and market change. It is impressive to note the acceleration of technology in these past ten years and imagine the ones to come; food tech has followed this same path, and of course, the questions our colleagues researching on technology ask for farming are valid for food technologies, specifically the one about data ownership, even more when it comes to the technologies employed at the household level.

Technology and Transition

For sustainability scientists, we are in the age of transition (e.g., Markard et al. 2012; de Haan and Rotmans 2011), specifically a transition to sustainability understood as "a fundamental transformation towards more sustainable modes of production and consumption" (Markard et al. 2012:955). The main aspect to grasp is the non-defined situation in which the ecological imperatives undermine the very existence of planet Earth; in fact "depletion of natural resources, air pollution and greenhouse gases emissions, nuclear risks, uncertainties related to short- and long- term security of supplies, and energy poverty" together with issues about water, extreme events, micro-pollutants, transportation sector issues aggravated by congestion, fossil fuel depletion, and CO_2 emissions (Markard et al. 2012: 955) have created a fragmented situation in which a new transformed stage has not been reached yet. Food-related fields of production, transformation, distribution, and consumption contribute massively to the issues of (non-)sustainability, because they touch upon many aspects of our lives. New or re-discovered crops are becoming available to global markets and this will change technology in turn. Agri-food is complex, as complex are modern societies. Such complexity is due to the many actors and many activities, coexisting or being more and more stratified. If ancient societies and agricultural practices can be considered as relatively easier to understand, modern ones are characterized by a high level of complexity because of a further element, that is, acceleration. Rosa (2013) talk about social acceleration to explain the tendency implicit in our modern societies to go faster as a result of technological acceleration and more demanding working and social conditions; this has implications on both material and immaterial aspects of our lives as individuals and as groups, as we get trapped into a loop that reinforces itself. Technology plays the central role in this sustainability as well as transition issue, because it is the engine and the outcome of this acceleration, and this is also the reason why we focus on transition as the hallmark of the past years. We have tried to keep track of our daily readings, just in newspapers and magazines, in which the words "technology" or "data" have appeared. It is basically impossible to document it, as it has become a pervasive topic and it would be a sterile exercise to report it here. It is pervasive, embedded in the very fabric of contemporary neo-liberal society (Smith Pfister and Yang 2018). In the Global North, basically, everybody has experienced the change of our private and professional lives to the point of disruption: technology is not any longer a matter of useful tools which complement our lives, but it has become impossible to work, operate, produce, and relate to each other without it or in a pessimistic view not even to think outside of it. For Zuboff (2019), this is the age of surveillance capitalism, in which technology and data usage have reached a level of alarm for citizenship (for a critical rebuttal on the basis of political economy analysis, see Morozov 2019); for some other commentators, we have reached a critical point in which a serious debate is necessary for preserving the very fabric of humanity, from biotechnology (Harari 2016) to brain and consequent modification (Harari 2018) through biology and neurology mining.

Food Tech Transitions

What makes it a transition age is the potential that technology has now reached and the pervasive role it plays, which forces the debate to exit the mere sectorial boundaries and enter the public arena, likely paving the way to a restructuring of the industry. In fact, some of the issues that affect agri-food, first, and society, second, are relevant for food technologies too, among which we have selected the following as they are present in the literature in a growing way:

(i)   The ecological imperative that forced a change in food production in the light of the advocated sustainable turn (e.g., Friedmann 2017) will affect the industry in a radical way, not much in terms of the longed agronomic revolution but most likely for the necessary adaptation of the needs of the same industry to ecology dictates, instead of the other way around; although the industry has moved steadily into sustainability, a significant change is not yet accomplished because it would undermine the very stability of the industrial system; therefore, a long reaction time is to be expected despite the quick character of recent techno-revolutions.

(ii) The food security imperative and the issue of food waste have been adopted by some food technology advocates to enhance the social legitimization needed (e.g., Nair 2016; Council and Petch 2015) but have to be yet fully addressed from a more analytical perspective. Both resonate strongly for a part of the audience they are intended for but cannot address some of the core issues at the basis of each, such as the still pervasive underlying productivist logic (Rosin 2013).

(iii) The shift of emphasis from production to consumption (Goodman and Dupuis 2002) is translated in the food industry by a shift to the household level (Truninger 2013) and professional catering, such as fine-dining, of those tools and techniques previously reserved to the core of the industry, which might cause a permanent alteration of meal patterns toward full digitalization of meal creation; this might have strong impacts at the nutritional level, might be hindered by cultural stigma for some ingredients and food, would reverse the hyper-diversification of production and specialization of some niche markets, with consequences for both production and distribution.

(iv) Altered global geopolitics and the social stratification (Bremmer 2018), especially in a period characterized by strong migration flows (UNHCR 2018), will likely be translated in an uneven development of food technology and its adoption, reflecting the already existent divide between macro-geographic areas and cultures and reinforcing much of the inequalities and limitations that the transition age requests instead to eliminate.

(v)   The steady move toward extreme digitalization (Ciuriak and Ptashkina 2018) extended to basically all the realms of life has already starting showing its limitations in terms of data access and privacy and seems to be stumbling upon some more wider ethical issues, such as integrity and trust, or compliance with governance; these have to be addressed, including the role of governing bodies in providing regulation for non-directly related aspects.

Organization of the Book

Consequently, the book is divided into two main sections, for which we propose a relevant perspective: Part One includes chapters that provide an overview of the changes of technology in production and in the processing industry and its (often uneven) advancements and the related ecological and production issues. Part Two includes chapters that address the more exquisite social science questions in terms of production and consumption, including also food-related technologies such as apps and social media, issues of societal change such as migration and the role of corporations in helping with transition. Also, we have excluded from our research new technologies such as blockchain for enabling traceability, as this pertains more to other sectors of agri-food while being central for issues of provenance.

Specifically, the first part is dedicated to the production issues in crop science and includes chapters that embrace a more agronomic and wider agricultural perspective, questioning the suitability and adaptation of existing plants and resources and novel food technologies and how they adapt to current trends in consumption and nutrition.

In Chap. 1, Khajehei, Piatti, and Graeff-Hönninger open the book through a historical overview that embraces social issues and set the questions that characterize this first part of the book: what are the food technologies we refer to, how they are situated in between production and consumption, and what are the challenges food technologies will have to face soon to be accepted. The authors embark in a historical journey following some social sciences theories, namely, risk society (Beck 1992) and food regime theory (Friedmann and McMichael 1989), to offer apt theoretical tools to correctly assess what kind of difficulties food technologies advocates and supporters will face. The authors interspersed the historical periodization offered by food regime theory with insights provided by risk theory, highlighting the difficulties in acceptance related to the political and social situation of each period.

In Chap. 2, Niakousari, Hedayati, Tahsiri, and Mirzaee propose an overview of the food industry and its advancement. This chapter is focused on the technological advancement in the food industry. It discusses the advantages and disadvantages of novel and emerging processing technologies, such as high pressure processing, pulsed electric field, microwave heating, radio frequency heating, radiation, infrared heating, ohmic heating, ozone, supercritical CO_2, etc., to be used for food processes and serves the aim of sterilization and pasteurization, degradation of toxins, modification of hydrocolloids, removal of antibiotics, reduction of insects, peeling, extraction, cooking, blanching, drying, thawing, tempering, concentration, etc. This chapter discusses how these novel technologies can improve the availability and quality of food products while being more fast, energy-effective, and eco-friendly in comparison to conventional heat treatments.

In Chap. 3, Trierweiler and Weinert propose that appropriate post harvest management and treatment of food products, in particular fruits and vegetables, are crucial for effective and efficient use of yield produced in terms of their physicochemical attributes including chemical, nutritional, and sensory characteristics and reduce the post harvest losses. Two of the most important key factors in this regard are temperature and the composition of atmosphere around the harvested fruits and vegetables during transport and storage time, as they affect the respiration rate of fruits and vegetables after the harvest. However, appropriate postharvest storage and transport is not always available specially in developing countries. Therefore, postharvest treatments such as hot water treatment, fermentation, and controlled and modified atmosphere packaging have been used to preserve the quality of fruits and vegetables with the aim to enhance the status of food security in such regions. This chapter primarily explains the methods for determination of quality of fruit and vegetable based on their different measurable physicochemical factors, the optimal cold storage and transport conditions, as well as the process parameters for post harvest treatment of various fruits and vegetables using hot water treatment, fermentation, and controlled and modified atmosphere packaging and UV-C treatment.

In Chap. 4, Acuña-Gutiérrez, Campos-Boza, Hernández-Pridybailo, and Jiménez detail the nutritional and industrial relevance of particular neotropical pseudo-cereals. This chapter discusses briefly the origins, traditional importance, nutritional relevance, and attributes of neotropical pseudo-cereals, namely, common bean, amaranth, quinoa, chia, chan, jicaro seeds, ojoche, and Andean lupine, which are gaining more and more attention of the consumers. Their potential to be included in a diet of specific groups of consumers, such as those with food allergies and chronic diseases such as celiac disease, is explained as these commodities promise to be healthy and nutritionally important and as such fit the expectations of modern consumers. The chapter provides a revision of their application in common food products and the availability of food products designed using these neotropical pseudo-cereals which have a huge potential and are among underutilized crops.

Chapter 5, by Graeff-Hönninger and Khajehei, closes the first part of this book, wrapping the insights offered by previous contributions and focusing on one of the current trends in the food production and consumption chain, the so-called super-foods, specifically on moringa, quinoa, chia, and yacón. This chapter is at the cross-road between the production field and nutrition issues. In fact, although there is no fixed definition for the term superfood, the term reflects the high nutrient content (e.g., antioxidants, vitamins, and minerals) in food products in general. The authors point to the emergence of low/no fat and low/no sugar products to meet the demand for food products which are healthy and/or health-promoting. The emergence of "all natural," "free from," and "no added" next to the trend for new diets such as "all-raw," "free from," and "vegan" represents the opportunity to incorporate the super-foods into the food products in such food chains. In reviewing the potentials of few superfoods to be used to new food designed for meeting specific diet requirement (e.g., gluten-free, low sugar, vegetarian, and vegan), the authors note how adaptation of super foods coming from different parts of the world from primary production to the last designed food products may help not only to benefit from their health-promoting aspect but also to enrich and revive the lost diversity of local crops. This chapter discusses the roots of consumers' behavior and food choices and their demand for superfood as well as the challenges that the current food system faces to meet such demand.

Part Two starts from this last point on food trends and shifts the focus on mainly sociological issues. This second part of the book opens with Chap. 6, which takes us into the micro-specificities of food technology to provide a reflection on the change of production patterns. Skartsaris and Piatti reflect on additive manufacturing and its application in the context of meal creation as a chance to look into production relations. The authors focus on 3D food printers and discuss how the claims of this technology's advocates extend to the modification of established production para-digms. The overenthusiastic series of claims that permeate this (and other) food technologies is discussed on the basis of post-Fordist paradigms; this results in the assessment of those claims that confirms how much production is still entrenched in unresolved issues that need to harness both internal and external forces and address change not on the basis of futuristic claims but on the basis of contemporary issues, which remain more problematic than envisioned by technology enthusiasts.

Enthusiasm is one of the terms employed in the next chapter, together with greening/sustainability and hedonism as characteristics of modern consumption. Chapter 7, ideally the companion of the previous one, focuses on food consumption, as we argue that a continuum between these two spheres help in addressing the transition character we refer to in this book. In this chapter, Piatti and Khajehei have searched the food-related landscape to make sense of the changes in food consumption. As some scholars (e.g., Warde 2015) have argued that food-related habits are paradigmatic of changes in consumption, the authors focus on food consumption and combine their research on some emerging trends which reflect the ecological, nutritional, health-based imperatives we have become familiar with.

In Chap. 8, Al-Sayed asks how will food technologies contribute to some of the most pressing issues that interest European and Middle East countries, that is, migration. Al-Sayed documents the migration flows of Syrian people fleeing their country, devastated by war, and arriving in Germany. The pressure posed by these migration flows, which demographics and social stratification are quite different compared to previous migration flows, provides the chance to reflect on the role of technologies, specifically in the context of food security and safety. The embedded role of technologies constitutes a given for both our societies and the migrants; as our secure societies are forced to change in response to the presence of different social groups of migrants, it is still to be understood whether technologies, applied to our everyday life in the context of food supply and consumption, can have a role for inclusion or will impact on further fragmentation.

Chapter 9 ideally takes the questions posed by Al-Sayed in the previous chapter and takes us back to the specificities of transition, discussing the role of corporates in the age of transition. Manning has started from the concepts disseminated throughout the whole book and has reflected on the role that corporations play in transitioning to more sustainable systems. Manning has employed the concept of regimes, suggested in the opening chapter of the book, being at the core of food provisioning, to highlight the structure existing in food provisioning; she has then reflected on the constellation of players, whether they are humans or non humans, which participate in the making of food provisioning and which push for sustainable and ecological relations. This screening drove her to analyze what is central for transitions as a dynamic moment in which all players are bound to each other and co-constitute the reality of societies and markets. Transition is a matter of actions and language employed, for which truth and related values are paradigmatic. Manning arrives at the heart of transition as a matter of social relations in which corporate responsibility is paramount and reminds us that the multiplicity of factors has to coalesce if we are to face the challenges of transition.

Stuttgart, Germany   Cinzia Piatti
                                                                                    Forough Khajehei
                                                                          Simone Graeff-Hönninger

References

Beck, U. (1992). Risk society: Towards a new modernity. (M. Ritter, Trans.). London: Sage.
Bremmer, I. (2018). *Us vs. them: The failure of globalism*. New York: Portfolio/Penguin.
Bryant, A., Bush, L., & Wilk, R. (2013). The history of globalization and the food supply. In A. Murcott, W. Belasco, & P. Jackson (Eds.), *The handbook of food research* (pp. 34–49). London: Bloomsbury.
Buttel, F. H., & Newby, H. (1980). *The rural sociology of the advanced societies: Critical perspectives*. Montclair: Allanheld, Osmun & Publishers Inc.
Campbell, H. (2012). Let us eat cake? Historically reframing the problem of world hunger and its purported solutions. In C. Rosin, P. Stock, & H. Campbell (Eds.), *Food systems failure, the global food crisis and the future of agriculture* (pp. 30–45). London: Earthscan.
Campbell, H. (2016). In the long run, will we be fed? *Agriculture and Human Values, 33*(1), 215–223.
Carolan, M. (2012). *The sociology of food and agriculture*. London: Routledge.
Carolan, M. (2017). Publicising food: Big data, precision agriculture, and co-experimental techniques of addition. *Sociologia Ruralis, 57*(2), 135–154.

Ciuriak, D., & Ptashkina, M. (2018). *The digital transformation and the transformation of international trade. RTA exchange.* Geneva: International Centre for Trade and Sustainable Development (ICTSD) and the Inter-American Development Bank (IDB). www.rtaexchange. org/.

Council, A., & Petch, M. (2015). *Future food: How cutting edge technology & 3D printing will change the way you eat.* Tumwater: Gyges 3DCOM, LLC.

de Haan, J. H., & Rotmans, J. (2011). Patterns in transitions: Understanding complex chains of change. *Technological Forecasting and Social Change, 78*(1), 90–102.

Floros, J.D., Newsome, R., Fisher, W., Barbosa-Cánovas, G.V., Chen, H., Dunne, C.P., German, J.B., Hall, R.L., Heldman, D.R., Karwe, M.V., & Knabel, S.J. (2010). Feeding the world today and tomorrow: the importance of food science and technology. *Comprehensive Reviews in Food Science and Food Safety, 9*(5), 572–599.

Fraser, A. (2018). Land grab/data grab: Precision agriculture and its new horizons. *The Journal of Peasant Studies,* 1–20.

Friedmann, H. (2017). Towards a natural history of foodgetting. *Sociologia Ruralis, 57*(2), 245–264.

Friedmann, H., & McMichael, P. (1989). Agriculture and the state system: The rise and decline of national agricultures, 1870 to the present. *Sociologia Ruralis, 29*(2), 93–117.Goodman, D., & DuPuis, E. M. (2002). Knowing food and growing food: Beyond the production–consumption debate in the sociology of agriculture. *Sociologia Ruralis, 42*(1), 5–22.

Harari, Y. N. (2016). *Homo Deus: A brief history of tomorrow.* Harper Collins.

Harari, Y. N. (2018). *21 lessons for the 21st century.* London: Jonathan Cape.

Helmond, A. (2015). The platformization of the web: Making web data platform ready. *Social Media+Society, 1*(2), 1–11. https://doi.org/10.1177/2056305115603080.

Kallinikos, J. (2006). *The consequences of information: Institutional implications of technological change.* Cheltenham: Edward Elgar Publishing.

Legun, K. (2016). Tiny trees for trendy produce: Dwarfing technologies as assemblage actors in orchard economies. *Geoforum, 65,* 314–322.

Markard, J., Raven, R., & Truffer, B. (2012). Sustainability transitions: An emerging field of research and its prospects. *Research Policy, 41*(6), 955–967.

Marsden, T. (1989). Restructuring rurality: From order to disorder in agrarian political economy. *Sociologia Ruralis, 29*(3/4), 312–317.

Michalopoulos, S. (2015). Europe entering the era of 'precision agriculture.' EurActiv.com23October Available online at http://www.euractiv.com/sections/innovation-feeding-world/europe-entering-era-precision-agriculture-318794. Accessed 18 Oct 2018.

Morozov, E. (2019). Capitalism's new clothes. Available online at https://thebaffler.com/latest/capitalisms-new-clothes-morozov. Accessed 5 Feb 2019.

Nair, T. (2016). *3-D printing for food security: Providing the future nutritious meal* (RSIS Commentary 273). Singapore: Nanyang Technological University.

Peters, B. (Ed.). (2016). *Digital keywords: A vocabulary of information society and culture.* Princeton: Princeton University Press.

Pilcher, J. M. (2016). *Food in world history.* New York: Routledge Press.

Rosa, H. (2013). *Social acceleration: A new theory of modernity.* New York: Columbia University Press.

Rosin, C. (2013). Food security and the justification of productivism in New Zealand. *Journal of Rural Studies, 29,* 50–58.

Silva, C., de Moraes, M., & Molin, J. (2011). Adoption and use of precision agriculture technologies in the sugarcane industry of Sao Paulo state, Brazil. *Precision Agriculture, 12*(1), 67–81.

Smith Pfister, D., & Yang, M. (2018). Five theses on technoliberalism and the networked public sphere. *Communication and the Public, 3*(3), 247–262.

Stewart, G. F., & Amerine, M. A. (2012). *Introduction to food science and technology.* New York: Academic Press, Inc.

Truninger, M. (2013). The historical development of industrial and domestic food technologies. In A. Murcott, W. Belasco, & P. Jackson (Eds.), *The handbook of food research* (pp. 82–108). London: Bloomsbury.

UNHCR. (2018). Forced displacement at record 68.5 million. http://www.unhcr.org/news/stories/2018/6/5b222c494/forced-displacement-record-685-million.html. Accessed 18 Nov 2018.

van Dijck, J. (2014). Datafication, dataism and dataveillance: Big data between scientific paradigm and ideology. *Surveillance & Society, 12*(2), 197–208.

Warde, A. (2015). The sociology of consumption: Its recent development. *Annual Review of Sociology, 41*, 117–134.

Wolf, S., & Buttel, F. H. (1996). The political economy of precision farming. *American Journal of Agricultural Economics, 78*(5), 1269–1274.

Wolf, S., & Wood, S. (1997). Precision farming: Environmental legitimation, commodification of information, and industrial coordination. *Rural Sociology, 62*(2), 180–206.

World Economic Forum. (2017). Shaping the future of global food systems: A scenarios analysis. http://www3.weforum.org/docs/IP/2016/NVA/WEF_FSA_FutureofGlobalFoodSystems.pdf.

Wrangham, R. (2009). Catching fire: how cooking made us human. New York: Basic Books.

Zuboff, S. (2019). *The age of surveillance capitalism: The fight for a human future at the new frontier of power*. New York: Public Affairs Hachette Book Group.

Part I
Food Tech, Raw Materials and Trends

Chapter 1
Novel Food Technologies and Their Acceptance

Forough Khajehei, Cinzia Piatti, and Simone Graeff-Hönninger

1.1 Introduction

When we were confronted with the idea of writing a book on food tech transitions we had to revise what it means to make sense of a particular subject; we came from different backgrounds – one of us from food technology, one from social science and one from crop science – and there couldn't be any taken-for-granted assumptions if we were to build a common ground for working together. Each had a different perspective but we agreed that we had to reflect on the challenges posed in front of us not only from a technical and sectorial perspective, and try to be as inclusive as possible because, on the basis of our academic experience, only the multifaceted contributions can deliver a significant tool for readers. It was clear that food technology is such a comprehensive term that can even be traced back till the scientific revolutions of 1500–1600s, but we all agreed that it was in more recent times that food tech has developed to the point of posing now central questions for our jobs, our role as consumers, our citizenship. It seems to us that two main challenges currently stand out among the many and they are interrelated: one relates to sustainability, to the very existence of humanity in its ecological context, about which food security is central – and still informs a strong social debate because of the alarming numbers provided by the FAO[1] each year – and therefore constitutes the second challenge. In the realization of the transition challenges of this second decade of the

---

[1] The United Nations Food and Agricultural Organization released the projected numbers on malnutrition and food insecurity for 2017, surprisingly on the rise, with of 815 million ca. people going hungry regularly.

F. Khajehei (✉) · S. Graeff-Hönninger
Institute of Crop Science, University of Hohenheim, Stuttgart, Germany
e-mail: f.khajehei@uni-hohenheim.de

C. Piatti
Department of Societal Transition and Agriculture, University of Hohenheim, Stuttgart, Germany

© Springer Nature Switzerland AG 2019
C. Piatti et al. (eds.), *Food Tech Transitions*,
https://doi.org/10.1007/978-3-030-21059-5_1

century, we agree that technologies will continue to be central and pivotal, because of their role in our lives and how we relate to them, and will constitute more and more the main axes around which agricultural production will revolve. Whether technology is value free or not is of course a huge on-going debate upon which we cannot focus in this brief space; suffice it to say, though, that before deliberating whether these technologies will have an environmental- or social- heavy impact we have to also be aware that how much of what we accept of technology and on what basis, depends on the historical period we analyse, which of course will concur in our assessment of the impact of technology itself. Different historical periods have seen a wider or more restricted acceptance of technology in our lives because they were characterised by different sensitivities. Whether the technologies employed in the food industry are going to be accepted or not depend on what sort of sensitivity is part of the social, economic, political and historical context of one period and one specific people. How we can make sense of all of this from an agri-food perspective is the goal of this chapter. In here we want to provide some composite reflections on the role and acceptance of food technology in the context of the two main challenges in agri-food (and beyond). For doing this, we employ the risk society theory offered by Beck (1992) to propose why technologies acceptance are welcomed or opposed in large social contexts, specifically from the past mid-century to understand the cultural turns happening in specific periods of uncertainty, as the one we live in. We do so following the convincing narration in the theory of food regimes as advanced by Friedmann and McMichael (1989), that proposes the existence of different periods of stability, crisis and regulation in food provisioning; in this periodization food security (declined differently in different time periods) provided a strong social justification for organizing production along industrial forms, of which the role of technologies is paramount, and consequently transforming consumption too. The strength of the theory is in the meticulous analysis of historical and cultural events, and in making sense of the role of actors and politics in the unfolding of each regime. The most recent strands of the theory specifically focus on the ecological imperatives characterising our age starting from the millennium, assigning a position and role to the juxtaposing forces on the stage (namely, corporations and social movements) and analysing specific trends, of which food waste has emerged as a (by-)product (pun intended) of the period after World War II. Aligning our argument with the periodization proposed by classic food regime theory, we want to situate food technologies in a manner in which their acceptance emerges clearly as both a challenge and the result of specific historical, political and cultural arrangements. We will start with food technologies in the context of food security, followed by an overlook on food regimes, in which the food tech development will be highlighted, in order to arrive to contemporary issues and novel food technologies.

1.2   Food Security and Food Technologies

According to the definition introduced by the FAO, "food security exists when all people, at all times, have physical and economic access to sufficient safe and nutritious food to meet their dietary needs and food preferences for a healthy and active life" (FAO 1996). Arguably, then, food security is a general term which emphasizes security in nutrition, the undeniable importance of food safety assurance, environmental and ethical issues (Karunasagar and Karunasagar 2016). The prediction that the world population will reach more than 9 billion by 2050 leaves no wonder about the fact that developing strategies to assure the global food security has attracted the attention of scientists from various backgrounds (Godfray et al. 2010; Karunasagar and Karunasagar 2016). Connections can be drawn between global food security issues and other global problems such as the sustainable management of rising demand for resources among which energy and water are the most important ones, together with climate change, and the progressive increase of world population before it stabilizes (Beddington 2010; Rosegrant and Cline 2003).

It has been stated that the solution to meet the food demand for the next decades might not be only by boosting the production output in primary production of food materials (Augustin et al. 2016). Optimizing the food processing system in the post-harvest end of food production chains in accordance with energy consumption, nutritional quality, yield of final products, and application of waste of food processing in other sectors (e.g. biofuel production, textile industry, chemical industry) or in development of value-added products can be influential too in improving food security by responding to contemporary sustainability issues such as energy crisis, malnutrition and waste management in post-harvest sectors (Augustin et al. 2016; Beddington 2010; Godfray et al. 2010). In addition, megatrends in the world have a foremost impact on the design of new foods product and the technologies which are used to produce them (Augustin et al. 2016; Hajkowicz 2015). Consequently, the perception of consumers toward food products and factors which will define the consumption habits and consumer acceptance should be determined in order to secure the success of these same new food products in the market (Augustin et al. 2016). In this sense, the development of food technologies in managing food products which will be accepted or rejected - and therefore might have an impact on food security- is paramount. This is why we focus on food processing technologies, as on one hand environment friendly and sustainable products constitute the new source of economic reward and help placate the modern consumers' anxiety over 'healthy and natural' food; and on the other hand can also provide viable fixes for meeting food security in both developed and developing countries.

In a wide way, food processing can be referred to any change to raw food material before its consumption (Floros et al. 2010). Such changes can impose negative effects to the food product by reducing the nutritional value because of the destruction of nutritional compounds. However, the benefits of food processing should not be neglected (Weaver et al. 2014). Food processing is essential to make the food consumable, increase the shelf life, enhance the bioavailability of critical nutrients in food, and destroy the toxic ingredient of food material (Van Boekel et al. 2010). Seasonality and perishable nature of food materials made the food processing a key factor to secure the food demands around the globe. However, as results of some factors including energy crisis, environmental impacts and nutritional losses during food processing by means of conventional technologies, research has focused on developing new techniques which enhance the food production chain by using sustainable energy while having less impact on environment and initial nutritional characteristics of raw food materials (Augustin et al. 2016; Pereira and Vicente 2010; Van Boekel et al. 2010; Van der Goot et al. 2016; Weaver et al. 2014). The interdisciplinary cooperation between pre-harvest and post-harvest sides of food production systems by taking advantage of novel technologies has been introduced as an effective way to make available to the population a diet that provides them with sufficient energy and nutrition besides satisfying environmental and ethical values (Augustin et al. 2016; Karunasagar and Karunasagar 2016). Food security, from a food industry perspective, would therefore emerge as the result of specific combinations of technological advancement made possible by social acceptance, political organization and market operations. In this regard, research and advancement must take a holistic and multidimensional approach taking all aspects of food production and consumption chain including cultural, economic, environmental, political, social and technological aspects combined into account. For that, an analysis of changes that food production and consumption has gone through and the present situation is essential.

1.3   Food Regime Theory and Food Technology

The 'Food Regime' theory as elaborated by Friedmann and McMichael (1989) was firstly introduced to explain the changes in economic and political characteristics of food systems in a particular period of history (Friedmann 2009; McMichael 2009) in which both food security and technological advancement played an important role. As such, it serves as a tool to identify the social, political and economic roots of success or failure of a functional food system for a period of time in history (Friedmann 2009; Sage 2013). Friedmann and McMichael posited the existence of a first (1870–1914 ca.) and a second (1945–1973 ca.) food regime, which have been then examined and used by several researchers to analyse food provisioning and their different aspects historically, economically and politically (Dixon 2009; Campbell et al. 2017). The theory later develops contending the existence of different regimes following the historical development of capitalist systems, upon which the US one is central for establishing much of the western world perspectives and practices globally. In particular, the role of the industry and food technology is paramount in both regimes but assumes a pivotal role during the second regime and in the aftermath of its crisis, until modern day. Campbell (2012) proposes that at the basis of the development of the second food regime is food security, which provided a social legitimation of food production organized around industrial patterns, of which food technology was essential. In the later development of the food regime concept, for which different food regime authors have different theories and names, technologies disappear from main view, and political and legal apparatuses are more evident but keep acting at an incredible pace. We hope to contribute to the re-visibility of them with this food regime theory periodization of food technologies.

1.4 Food Tech During First Food Regime

The First Food Regime was used to determine the properties of food provisioning between 1870 and 1914 and addressed its characteristics as a function of hegemony in the world (McMichael 2009). In this regard, this regime has been used to describe the causes and effects of British hegemony on food production under colonial empire configuration. During this period, the main inventions and discoveries of the Industrial Revolution were employed and maximised, mostly in terms of transporting food and raw material from tropical and temperate settler colonies to feed the rising population of working class in European countries such as Great Britain. Some tropical food commodities which shaped the diet of industrial workers in UK and Europe were vegetable oils, tea, coffee, sugar and bananas made their entrance on the diets of Western people and stayed since then. Such move contributed to stabilize the status of food security in the growing industrial Europe by importing wheat and meat, which became staple foods for industrial workers in Europe; the settler states, on their side, imported manufactured commodities, labour and some forms of capitals from European countries. In this regime, basically, distance becomes a main characteristic in the food production and consumption continuum. Friedmann and McMichael (1989) propose that food production, which has the main characteristic of durability in this period, was relocated in the settler states where the advancement in technology contributed to adaptation to cheaper forms of agricultural production. The main technologies employed here are those related to transport with the advent of railways and the first big shipments of refrigerated meat, and those related to mechanization of work in the field, namely ploughing machines, thresher, reapers, water- or wind- energy motioned mills and animal-led equipment being at the basis of a somatic energy regime (Derry and Williams 1960); in fact mechanized harvesting helped the settler agricultural system to overcome the issue of the shortage of labour. In terms of properly said food technologies, the main forms of food transformation already in use were perfected and use of fossil fuels for making engines work were employed. These technologies, although quite basic for us in the 21st century, had the capacity to revolutionize production and processing operations; therefore from a sociological perspective acceptance was arguably determined by the time freeing and human labour saving, something most longed for. At the household level the common tools used in kitchen (and cuisine) in this period are the maximum form of help household keepers (usually women) have, and basic forms of knowledge passed on for generations such as those for food preservation and storage (such as curing, salting, smoking, brining, pickling; Huang 2000) are still the most important guarantee of food availability. Also raw forms of modern technologies based on the the then-beginning chemical industry (Truninger 2013) make their appearance, for instance canning around 1850s, and the discoveries of Appert in terms of vacuum sealed containers and Pasteur on the role of bacteria a decade later (Thorne 1986) which enhanced the canning industry techniques. In particular, the meat canning's history indeed illustrates Friedmann and McMichael's observation of the relations between settler states and empires, as this branch of the canning industry developed in Australia and South America enabling 'the export of cheap meat to feed the working classes of European industrial cities' (Truninger 2013: 84). Lastly, refrigeration became the indispensable partner for allowing the industry to progress from durable to perishable foods: the industry of the cold chain developed in reaction to negative perception (such as food not being fresh or at the risk of poisoning) of food being stored using natural ice, as was the case for products travelling from Australia, South America and South Africa (Teuteberg 1995), and paved the way to the introduction of artificially-made cold, 'revolutionizing the organization and scale of production, storage, distribution and ultimately consumption' (Truninger 2013:84). For Friedmann and McMichael (1989) this state of the affairs remains stable until the advent of World War I, which causes enormous disruptions in the way food provisioning is operated. The time gap between the emergence of a second regime is referred to as an experimental and 'chaotic' era (1914–1947) during which the main features of the first food regime evolved into the second one to meet the demand of world population for food (Friedmann 2009; Sage 2013). This transitional period may be characterized by several remarkable historical turning points including World War I, the Great Depression, followed by World War II. A second regime emerged and stabilized between 1943 and 1973 ca., during which food production and consumption were dramatically transformed under the transition of colonial empires into capitalist nation-states configuration (Campbell et al. 2017), the most important of which was the United States of America (Friedmann 2009), a country which has made technological advancement at the forefront of economic stability.

1.5   Food Tech During Second Food Regime

This second regime stabilized by taking advantage and pushing for more of the technological advancement that started during wartime; the tools and machinery then employed were forced into agricultural application and domestic agricultural production in both industrialized and non-industrialized countries since 1950s, the most famous and important outcome of which is known as the Green Revolution (GR; Campbell 2012). In fact, this second food regime owes its success to the investment in crop science, development of infrastructure and market, as well as policy support during the GR and post GR period until 1970s (Pingali 2012). The GR response to the growing population was through adaptation of high yield new breeds and varieties, application of fertilizers and mechanization of agriculture with the fundamental support of governments in terms of subsidies (Campbell 2012). GR led to remarkable shifts in productivity of the food supply chain, to significant production of specific staple crops (wheat, maize, rice) and contributed to significant change in diet, giving rise to a new system of food consumption (Sage 2013). The raise and stabilization of globalized fast food restaurants, big food industry, and retailers and eventually powerful cooperation operating and controlling the food system stemmed from the many domestic and international policies that were in place to support this wave of agricultural intensification. Such policies of food aid and distribution, particularly the policies of USA with reference to production supports and price stability system, followed the Marshall Plan and contributed to the political hegemony of the USA; this set of operations went hand in hand with the GR-created agricultural surpluses and helped stabilize the conditions typical of this second food regime. The foreign policies forced by US and later in UK and Europe accorded the flow of agricultural surplus and their processed food products in the form of food aids to developing countries and ultimately it constituted the beginning of the growth of agri-corporate (Campbell 2012). It is in this context that the main evolutions of food technologies as we know them today happen, and we contend that it happened on the basis of the role of agri-corporations, which relied massively on development of further technologies, as they respond to the need of mass production. There is a subtle observation to be made here: the social legitimation provided by food security we mentioned in a previous section is the outcome of the post-war compromise; as Campbell (2012) points, the experience of hunger during the war was paramount in allowing for the creation of common grounds for rebuilding Europe. It is clear that this sad experience, together with the brutalities of war, was the common element which cemented the international agreement for creating supranational institutions (necessary in a world constituted by nation-states instead of empires) that would implement plans to ensure peace and reconstruction. This would have been done enhancing international agreements, promoting cooperation and enhancing production through the means of technology, among which food production was paramount in the first years after the war. It is in fact in these years that a strong cooperation between political institutions and scientific ones begin, as the memories of war had pushed for hope in something neutral (or impersonal) enough which could ensure that human beings would not get back to the past atrocities, namely markets and science. It is here that we see the precedents of Beck's theory of risk society (1992, 1994); Beck proposed that throughout history from the starting point of industrial revolution, societies went through changes and modernization which brought them to this specific moment. German sociologist Beck (1999) believed that this period was one of immense trust in science and technology as a sign of modernity, of being different from the past because of the possibility to overcome uncertainty (represented by risks). The economic boom constitutes a period of exclusive trust in these institutions and the human intellect to deliver, through technology, solutions for production and consumption for the betterment of human living conditions, of which hunger reduction was essential; this would last 20 years ca., and would then transform again in unimagined ways.

Most importantly for our argument is that the second food regime represents a period of mass production and consumption of new and heavily processed food products (Pritchard 2009). First of all, the canning techniques mentioned before reached an important level of precision and industrialization; this was used at the beginning in the military and then, later, as a way of penetration of American culture and products in Europe (Thorne 1986). Secondly, refrigeration techniques advanced, not only at the industrial level but more importantly entering the households in a massive way only after World War II. For Freidberg (2009) this was made possible together with positive reinforcement of the importance of freshness and cold storage starting from World War I from governments, nutritional science and industry. Thirdly, a new diet based on widely versatile processed food commodities produced in the factories in this historical period, went global. It is in this time that some authors (such as Popkin 2003; Dixon 2009) locate the transformation of a plant-based diet into a diet enriched with animal-based foods, fats and oils, processed sugars and carbohydrates, for which scholars coined the term 'nutrition transitions' that characterises much of the change of diet during the second food regime and after. The nutritional transition has taken place in two phases. The first phase includes an increase in the diversity of diets, high meat and lightly processed products; and the second phase involved a creation of diets specific for different classes: a working/poor class with a diet based on the relatively cheap, high calorie and highly processed food products, and a wealthy class with a diverse diet containing expensive fruits and vegetables (Dixon 2009; Hawkes 2009).[2]

A final point has therefore to be made about the advancement in technologies, which made the functionality of the two food regimes possible (Friedmann and McMichael 1989). It has been pointed out that distance between place of production and consumption relying on durable food products has caused adverse environmental impacts and the failure of first and second food regimes (Campbell 2009; Friedmann and McNair 2008). Food commodities which were produced as a results of technical and scientific advances typical of the years 1930–1970s helped the production and distribution of food and diets around the world during both first and second food regime (Dixon 2009). One last thing we want to note and which is not highlighted in the theory proposed by Friedmann and McMichael (1989) is the role of the military industry during this period of time. Truninger (2013:85) maintains that food innovation was pushed by this complex, as its technologies were then strongly applied for commercial use, as is the case of microwave ovens and plastic gadgets for storing food (e.g. Tupperware) or food irradiation for preservation. Zachmann (2011), though, notices that the perception of such food and its potentially dramatic effects on health put at risk the trust and acceptance among consumers

---

[2] This second regime, Friedmann and McMichael (1989) propose, was stable until mid-1970s – although for some authors such as Pritchard (2009) even until the 1980s – when rising trade wars between US and EU as a results of increase in agricultural surplus in Europe on top of crisis in US agriculture challenged the hegemony of US in food production system. As results, food and agriculture, until then excluded from international trade agreements, were included in the multilateral trade system and ignited the regulation of a global of food politics (Pritchard 2009).

and even some manufacturers. All in all, these technical scientific advances not only were quite different from those occurring during the first food regime but they were different also from the beginning of the second food regime. In fact, if the years after the war were the years of escaping hunger, the next 10–20 years became the years of modernisation, because technology had delivered what was promised. But it is also the moment in which the first signs of a cultural change in the attitudes towards this modern food and the technological innovations that delivered it, emerge.

1.6 Beyond the Second Food Regime

Although the second food regime was successful for a period of time to stabilize the political and economic system by which the world worked, it entered as well a period of crisis subsequently to main historical changes, as happened to the previous regime and as might happen to subsequent (if any) regime. During the 1980s the first signs of a negative impact of GR and intensified agriculture started to come into view (Pritchard 2009). The high yield new varieties were in particular responsive to the external input such as fertilizers or pesticides and, together with irrigation in the field, were at the centre of high productivity achievements. However, in accordance with a lack of proper policies, the consequence of intensified agricultural system has its negative impacts such as soil degradation, over-consumption of water resources, and leakage of chemicals into the ecosystem on the environment (Pingali 2012). This has manifested in the shape of lower productivity in mid 1980s as a result of degradation of agricultural resources such as soil, water and diversity of crops (Pingali 2012). Hence, the consequences of agricultural intensification are among the root factors of some of the global challenges currently experienced, including the ecological issue. Moreover, some of the current global health issues including the rise of obesity and type II diabetes have stemmed from the new diet forms which were globalized during the second food regime. Again Beck (1992, 1994, 1999) explains that if modernisation of a society happens through the process of innovation, then our present society has gone through two distinctive phases. The first phase includes the stages during which the consequences of innovations existed. However, these did not concern the public and political conflicts were not emphasising on them. And still, when the negative consequences or hazards of industrial society emerged to surface and concern the public, they became the centre of political and private discourses. This is the moment when, according to Beck, an industrial society transforms into a risk society. In a risk society the conflicts over distribution of hazards (or 'bads') produced are layered on the disputes over the distribution of societal 'goods' such as income, social security and jobs. This may be introduced as the foundation of conflicts in the industrial society over how the risk or hazards of production and capital accumulation of commodities such as the risk of application of novel technologies, use of chemicals and chemical technologies, and concern over environment, may be addressed (Beck 1996). This will become increasingly important in the aftermath of the second food regime because such an over exposition to risks, without the safe net that science represented in the past, does not leave any escape, any alternative or offer any hope, because the saviour (technology) has transformed into the persecutor.

Now, if we have to assess the two regimes, the first and second food regime were functional when the global food relations were stabilized around the political and governance arrangement, trade trends, labour relations, farming systems, commodity complexes, consumer cultures; and were destabilized during the transitional periods or period of crisis (Campbell et al. 2017). Some researchers have identified the period from 1980s to the present time as a transitional period in terms of food regimes while others have proposed the emergence and existence of a third food regime by weighing in the ecological and cultural dynamics into account (ibid.). In this regard, the existence and emergence of a third food regime is highly debated. In agri-food circles, two main theoretical propositions have been credited: one named 'corporate industrial food regime' proposed by McMichael (2005), while other researches, including Friedmann, framed the current status of food relations under the name 'corporate-environmental food regime' as a possible third food regime in place (Burch and Lawrence 2009; Campbell 2009; Friedmann 2005; Holt Giménez and Shattuck 2011; McMichael 2009; Sage 2013).

The 'corporate industrial food regime' (McMichael 2005) would be a regime dominated by corporations and the whole political and legal apparatus for ensuring its domination; this structuration means that food technologies are heavily employed as foundational part of the working of the regime, specifically in terms of less labour presence and more and more automated, technology-based solutions for processing, transforming and creating meals. If this is the regime that currently exists, then it is inevitable to notice that it attempts to use green strategies for sustaining agri-food systems, for example by promoting production methods and systems which steer away from external inputs (e.g. chemical fertilizers and pesticides in the agricultural sector, and chemical additives in the food processing). It is to note that this move is proposed by McMichael as 'business as usual'. In fact, anything that is different from this industrial production organization is either subsumed or marginalised; in this regard, the so called 'alternative food networks' (Goodman and Goodman 2009), defined as networks of food production, distribution and consumption which try to escape the industrial logics and are so familiar to the average western consumer, are the proof of the ability of regimes to maintain their hegemony, because they pose no threat to a corporation-based regime (and likely will follow the same path of organic agriculture in becoming conventionalised). The existence of this regime exacerbates both the role of technologies as complementary with the evolution of such regimes, and the issues related to food security, as the strictures of this contemporary regime has its precedence in the colonial strictures typical of the previous regime, and on non-resilient agricultural and economic systems.

On the contrary, Friedmann (2005) and Campbell (2009) think that these same alternative food networks and the existence of other actors and trends (such as supermarkets, audits or organics) are evidence of the existence of a different configurations of relations, that is: a different regime named 'corporate environmental' exists in which market transactions are fewer and of lower intensity (and therefore weakening capital accumulation is subsequent, as suggested by Friedmann 2005). The emergence of alternative food networks and their growth would therefore be a signal showing a growth of a new line of food production and consumption chain for capital accumulation by placing reliance on the environmental movements and emphasising concepts of fair trade, natural food, promoting health of consumers, and counting in the values of animal welfare (Friedmann 2005). These networks are the opposite of technology-intensive systems as long as they are characterised by a local, re-spatialised and re-socialised nature. More importantly for our argument, usually consumers participating in some forms in these networks are technology-sceptics, as is the case for biotechnology and nanotechnology, which Truninger (2013) believes underwent the same negative perceptions that had accompanied the technologies employed during the 1960s. In addition, concerns for food security assume a different perspective, as food provisioning which is not trapped in industrial and capitalist production is less rigid, and as such shocks (environmental, political, social or economic) can be absorbed without hitting as in the past or concurring towards evolution (and possibly shock reduction).

1.7 Novel Food Technologies

In this contemporary context, again Beck's concepts are useful, and specifically his proposed idea of a reflexive modernity (1994) is helpful to grasp some of the main issues around technology acceptance. Risk society, in fact, posited a division between nature and society and an overreliance on rationalism and progress as hallmarks of modernity. We were left in the previous sections with an over-pessimistic view of modern societies as being far and distant from nature, and with the delusion of rational solutions; there is nowhere to turn to because main institutions have left individuals alone and in risky situations. Beck came to the conclusion that this dynamic of risk develops in a dialectic way so that a modernity unfolding this way has to transform, and it does so trough individual reactions and initiatives. In fact, new risks posited by technological advancement are real and are clearly perceived in a negative way, which means that people are not blind to the risks and limits of technological advancement, and have to confront them -as well as confront themselves- in a reflexive exercise. In fact, the rise of so called alternative food networks during the past two decades may be evidence of this, and of such strategies put in place mentioned above. Take the establishment and growing of food supply chains such as organic food production systems, fair trade and local production and consumption in this regard as a proof of the attempt to escape the capitalist logics (Friedmann 2005; Levidow 2015). These 're-born' consumers advocate for less food processing and target environment friendly and healthy food, and lean on alternative forms of production. It has to be noted, though, as Dixon (2009) does, that the industry is happy to provide them exactly what they ask for, with enormous investments in the nutrigenomics and functional foods.

It is exactly here that food techs fit in this contemporary structure of food provisioning, as they are the tools that allow the industry to take advantage of the major global trends of healthy and sustainable food; for making sense of this, we leave now aside the double focus on both industrial and household level, and revert our attention to the technologies available in the industry only, as this is where the big changes are currently happening, and because they are not available at the household level for obvious reasons. As it is discussed in Chap. 3, various novel non-thermal food processing technologies including pulsed electric fields (PEF), supercritical CO_2, high pressure processing (HPP), radiation, and ozone processing as well as novel thermal processing technologies such as microwave, ohmic heating (OH) and radio frequency (RF) heating have been developed and regarded as alternative to conventional heat treatments in recent years. Moreover, in Chap. 3 it is noted that these technologies may be used for different food processing such as pasteurization, sterilization, drying, peeling, cooking, or extraction for a wide range of food products, while production lines in different food industries have been profiting from their advantages. Furthermore, according to recently published results, such as Jermann et al. (2015), in Europe and North America HPP, PEF, MWH, and UV are the four novel food processing techniques already commercialized or having a good chance to become commercialized in 5–10 years time. This shows that novel food technologies are already taking their role in shaping and stabilizing the present/emerging food regime as they respond to the needs of the industry.

Accordingly, the potentials of novel food technologies to enhance the status of food security, in the context of the "corporate-environmental food regime" will be discussed in the following sections of this chapter. In fact, it can be argued that to maintain food stability, the new food regime referred to as 'corporate-environmental food regime' might be based on local and seasonal food production systems (Friedmann 1993; Friedmann and McNair 2008). In this regard, the use of the novel food technologies might improve the stability of local food production while imposing less environmental impact and less energy consumption to extend the shelf life of locally produced food products. In addition, taking the example of HPP, this technology not only reduces the energy consumption and is operational using green sustainable energy, but it also makes the microbial decontamination of food products possible by using high pressure, hence increasing the shelf life and safety of food products with less impact on nutritional factors - in comparison to conventional thermal process (Rendueles et al. 2011; Wang et al. 2016). Same advantages may be pointed out for the other novel food processing technologies. Therefore, beside production of safe food and reduction of losses in nutritional factors of the product comparing to conventional food processes, such novel technologies are empowered by green and sustainable energy (Jermann et al. 2015; Pereira and Vicente 2010; Sims et al. 2003). Moreover, novel thermal and non-thermal food processing technologies has been evaluated as energy and water saving while they may reduce the emission of food processing (Masanet et al. 2008). Consequently, they may be used to improve the efficiency of food production systems in the framework of "corporate-environmental food regime" and help stabilize the relationships occurring, in favor of the corporate side.

Over and above that, in accordance with the recent argument of Campbell et al. (2017), the re-visibilisation of food waste in the current food regime is another significant factor at stakes in the present time. Food waste was not an issue before, in the configuration of previous regimes, because it simply did not exist in the form we know it today; shortages made impossible to waste anything, actually on the contrary they are arguably what enabled food habits and traditions to be created. The authors maintain that the problem of food waste emerged to the surface of food regimes as a result of several watershed moments including the growing concerns over climate change, rise in oil prices, rise in food prices in 2008, 2011 and 2012 which contributed to public awareness and the creation of specific policies about the future of food supply chains and their sustainability. The authors pointed at the environmental management frameworks such as EU Landfill directive (1999/3/EC) and creation of Waste and Resources Action Programme (WRAP) in the form of a non-profit company. Furthermore, WRAP contributed remarkably to bringing the food waste issue into public attention initiating the campaign "Love Food Hate Waste" (2006). Later, other organizations such as FAO took action to quantify food waste and food losses by investigating the whole food chain. The results of such investigations explained that the food waste in the Global North is a problem with its roots in the retailing sector and along the consumption chain, while in the Global South it is mainly stemmed from postharvest losses in the food production system as an outcome of technological failure and lack of efficiency in organisation management (Campbell et al. 2017). The valorisation of food waste is one of the strategies that may be used to reduce the cost of food loss and food waste (Mirabella et al. 2014) and in this context it is foundational to use technologies which are environment friendly and 'green' for waste valorisation. In this regard, the potentials of various novel thermal and non-thermal processes have been investigated by many researchers. For example, for the extraction of valuable phytochemicals such as phenolic constitute of waste of fruit and vegetables, the use of supercritical carbon dioxide extraction, microwaves, ultrasound, pulsed electric fields and high pressure has been introduced as an environment friendly process, as these extraction methods do not require organic toxic solvents (Putniket al. 2017; Mirabella et al. 2014). In this regard, take the example of banana peels as a waste of banana production. Thirty percent of a ripe banana is its peel. Banana peels are a valuable source for pectin production, micronutrients to feed cattle and poultry, wine and ethanol production, biosorbants for detoxification of feed, production and utilization antioxidants to be used in nutraceuticals (Mohapatra et al. 2010). To revive the valuable compounds waste of food processing, the extraction of the valuable compounds with enhanced yield and energy consumption using novel technologies such as MWH is a good example (Jia et al. 2005; Qiu et al. 2010). Back to the example of banana peels, the utilization of banana peels for production of biogas and removal of heavy metals and radioactive minerals from wastewater can be named as a good example to show the potential of such by-product to enhance food security by improving environmental conditions, reduce fossil fuels dependence and resuscitate water resources (Oyewo et al. 2016; Wobiwo et al. 2017). Food waste and novel food processing technologies, in addition, may be linked in a further way, for example, the radiation of meat products is a non-thermal novel technique for their sterilization, although the irradiation of meat may affect its sensory quality according to the oxidative changes. Also, the study of Kannatt et al. (2005) showed the effective use of potato peel extract for retarding the peroxidative changes in radiated meat produced. Where does this leave us? Well, whether it is one or the other regime (or even none of them), it is undeniable that there is a coexistence of different and opposing trends which pull in different directions, corporates hegemony is strong and environmental issues are central. So, again, what is then the role of food technologies in this? Whether we embrace the understanding of current food provisioning as being corporate industrial or corporate environmental, one question emerges as to whether these food technologies can help in the making of more sustainable food regimes in which environmental and health related concerns are resolved. Although it is evident that the very formation and running of food regimes is characterised by technological innovations, the demand for more sustainable systems (instead of regimes) calls also upon the role of these technologies, and given the coexistence of both pro-technology organisms (corporations) and anti-technology ones (alternative food networks and the likes, to make a bold differentiation, although of course there is a good degree of acceptance almost in these networks) it is probably more helpful to make sense of them and see how they can adapt to one or the other, help running out of the strictures of food regimes, help in addressing food security and possibly contribute in the making of better systems.

1.8   Consumers' Attitude Toward Novel Food Technologies

Consumers' attitude and acceptance are crucial to make the commercialization of new technologies in food production chain possible, even more in the context of food regimes, as the regimes rely massively on legitimization, of which acceptance by the public is paramount. It has been noted that consumers are favouring food products which are minimally processed, contain fewer food additives, have less adverse impact on health, enhanced the health condition and produced sustainably (Bhaskaran et al. 2006; Jermann et al. 2015; Zink 1997). It has been stated that negative attitudes of consumers toward processed food products has its roots in consumers not trusting technologies, in the lack of knowledge about processing technologies, in counter-advertisement against food technology by food activists, in the concerns about heavy use of sugar and salt in processed food, and in the beliefs such as the lack of nutritional benefits in all food produced by big food companies (Floros et al. 2010; Williams and Nestle 2015). During the past two decades, a considerable number of studies has noted the guarded attitude of consumers toward novel food technologies (Cox and Evans 2008). It is worth discussing the roots of negative attitudes of consumers toward novel food technologies or in general novel technologies going back to Beck's risk society (1996) for a last rush. We have seen how after the strong belief in science and technology that characterize the second food regime we have moved into a period of distrust. The argument back then was that it is preferable to take some risks in a private and individual way that to trust science and technologies, empowered by the political bodies, that have caused the environmentally dangerous situation we find ourselves in. From a consumer attitude perspective, our society is a risk society and in a risk society the hazards of industrial society are dominant (Beck 1996). Hence, in the risk society the individuals are concerned and critical over the application of technologies and science which are perceived as the roots of global challenges. So, the critical view and negative perception of consumers toward application of novel technologies in food production and consumption chain may be explained by the nature of the risk society itself. The negative consequences of trusting science and technology during the second food regime is contemplated as the main origins of current global challenges. Moreover, the level of mistrust in the application and use of novel technologies for food processing and caution may vary among wide ranges of consumers, with regard to a wide range of novel technologies (Cox and Evans 2008), so that it is not unidirectional and has to be faced from various perspectives, such as the different consumers in different place, the differnt food items and the different technologies applied. In this regard, risk can be divided into different categories in accordance with consumers' perception of risk (Slovic 1987). For example, if the risk is voluntary or involuntary, observable or hidden, immediate or delayed, fatal or non-fatal; and according to the consumers' degree of control over risk, and the degree of science and information about the risk itself (Cardello 2003). In this respect, the risk of using novel technologies may be regarded as an involuntary risk that individuals are exposed to by consumption of food which may cause irreversible changes in food products and lead to unknown delayed health risk (Oser 1978). Take the example of consumers who are favouring the green and alternative food networks and are resistant to technologies and, accordingly, show greater levels of concern over the application of novel food technologies. Contextually, there exists another group of consumers who put their trust in the science and regulatory authorities of food production and consumption, and are willing to consume food products which are manufactured using novel food technologies (Bord and O'Connor 1990; Bruhn et al. 1986; Bruhn et al. 1996; Cardello 2003; Sparks and Shepherd 1994). This means that acceptance of novel technologies and consequently some food products depends significantly not only on the perceived benefits of them but also on their perceived risks, and that this coexistence will likely pull in different directions, juxtaposing one to the other. Furthermore, many consumers have limited knowledge related to the novel technologies, and for these consumers trust is an important element, actually paramount. However, based on the failures of the food industry during the second food regime to secure and sustain the trust of consumers, the public may not be at rest about values that dictate how food is produced at the industrial level and by means of novel food technologies. In this respect, one of the key roles during the second food regime belonged to the policies in place. The policies in place back then may be blamed to a great extent for the adverse consequences of GR and, later on, for the expansion of industrial food and fast foods. In this sense, the lack of trust in the information made available to the public by or through the food industry is understandable. Respectively, providing the information toward benefits of novel technologies to public through independent consumer organizations or scientist are more welcomed by the public and might induce more positive influence toward their acceptance (Siegrist 2008), but the industry has to be careful to avoid the perception of green washing. It is essential to present the correct and understandable information about the food produced by means of novel technologies to secure their acceptability by consumers (Liu and Lopez 2016) but even more to work towards reconstruction of trust. Some old issues, such as acceptance of GM for example, can re-emerge even stronger than before as it may be extended to other novel food technologies (Siegrist 2003, 2008).

1.9 Conclusion

Food security is indeed not only a global challenge of our time, but also an imperative to ensure a healthy human society (Augustin et al. 2016). Since prehistory time, humans started using food processing for extending the useful life of food materials and their safety as a part of primitive food production and consumption systems. Specifically, at this point of our history, the food production and consumption system is summoned to respond to a very complex demand for food. It is not only about fixing the hunger and/or nutritional inadequacy, but also about doing so with sustainable approaches to ensure the survival of our ecosystem as a whole. Therefore, food production is not only about the call to increase the production to meet the demand in consumption at any cost. Here, sustainability is a prerequisite to all sectors of food production and consumption system including the food processing and technologies. As discussed in this chapter (and as will be seen in Chap. 3), research and food techs development have been focusing on innovative food technologies to be able to adapt to the growing demand for sustainability in terms of energy and water consumption, preservation of nutritional value of food products, reduction of food waste and ensuring food safety. In addition, the existence of global megatrends (namely, 'more from less', 'planetary pushback', the 'silk highway', 'forever young', 'digital immersion', 'porous boundaries' as proposed by, e.g, Augustin et al. (2016) or Hajkowicz (2015)) will affect the way food products are designed and produced. Such trends are originated from political, economic, social and ecological situations which consequently will alter the lifestyle of the earth population. Hence, the food consumption patterns will also be influenced by such trends (Augustin et al. 2016; Hajkowicz 2015). The backbone of these mega trends is accommodating and coordinating the elements of innovation in terms of technologies, promoting health, protecting the environment and cultures, reducing waste, and enhancing consumers trust. In other words, the characteristics of the 'corporate-environmental food regime' is evidenced and reflected in these trend. For examples, in the context of 'more from less,' attempts to minimize the waste of food processing or use of waste of food processing in other industries may be made (Hajkowicz 2015). The novel food technologies may play a significant role in solidify such trends.

Although, according to information and knowledge available, the novel food technologies may be able to adequately respond to the requisites of sustainability and food security in the framework of a 'corporate-environmental food regime', the negative attitude of consumers and nature of risks are among the most significant factors standing in the way of benefiting from the novel food technologies to their fullest potentials. In their investigation over GM products and related fears, Campbell and Fitzgerald (2001:218) have noted that "fear over food technology is a complicated subject requiring further analytical thought and empirical investigation. [...] not all new technologies are subject to continues stigmatisation and not all scares are effective in reducing access to the stigmatised product within a society". Therefore, at the current time, one of the outstanding confrontations is to coordinate consumers' acceptance of the currently available novel technologies. Concerns have become evident to the public as a consequence of trusting innovations and technologies blindly during the second food regime, and mistrust of consumers in innovations and novel technologies are direct result of that. Consequently, in-depth studies into consumers' benefit and risk perception of novel technologies may be among the most decisive. This may result in development policies for productive communication of information to effectively gain the trust of consumers (Bearth and Siegrist 2016).

References

Augustin, M. A., Riley, M., Stockmann, R., Bennett, L., Kahl, A., Lockett, T., Osmond, M., Sanguansri, P., Stonehouse, W., Zajac, I., & Cobiac, L. (2016). Role of food processing in food and nutrition security. *Trends in Food Science & Technology, 56*, 115–125.

Bearth, A., & Siegrist, M. (2016). Are risk or benefit perceptions more important for public acceptance of innovative food technologies: A meta-analysis. *Trends in Food Science & Technology, 49*, 14–23.

Beck, U. (1992). *Risk society: Towards a new modernity* (trans: Ritter, M.). London: Sage.

Beck, U. (1994). The reinvention of politics: Towards a theory of reflexive modernization. In U. Beck, A. Giddens, & S. Lash (Eds.), *Reflexive modernization: Politics, tradition and aesthetics in the modern social order* (pp. 1–55). Cambridge, UK: Polity.

Beck, U. (1996). Environment, knowledge and indeterminacy: Beyond modernist ecology. *Risk, Environment and Modernity, 40*(27), 27–43.

Beck, U. (1999). *World risk society*. Malden: Polity.

Beddington, J. (2010). Food security: Contributions from science to a new and greener revolution. *Philosophical Transactions of the Royal Society of London B: Biological Sciences, 365*(1537), 61–71.

Bhaskaran, S., Polonsky, M., Cary, J., & Fernandez, S. (2006). Environmentally sustainable food production and marketing: Opportunity or hype? *British Food Journal, 108*(8), 677–690.

Bord, R. J., & O'Connor, R. E. (1990). Risk communication, knowledge, and attitudes: Explaining reactions to a technology perceived as risky. *Risk Analysis, 10*(4), 499–506.

Bruhn, C. M., Schutz, H. G., & Sommer, R. (1986). Attitude change toward food irradiation among conventional and alternative consumers. *Food Technology, 40*(12), 86–91.

Bruhn, C. M., Schutz, H. G., Johns, M. C., Lamp, C., Stanford, G., Steinbring, Y. J., & Wong, D. (1996). Consumer response to the use of lasers in food processing. *Dairy, Food and Environmental Sanitation: A Publication of the International Association of Milk, Food and Environmental Sanitarians, 16*(12), 810–816.

Burch, D., & Lawrence, G. (2009). Towards a third food regime: Behind the transformation. *Agriculture and Human Values, 26*(4), 267–279.

Campbell, H. (2009). Breaking new ground in food regime theory: Corporate environmentalism, ecological feedbacks and the 'food from somewhere' regime? *Agriculture and Human Values, 26*(4), 309–319.

Campbell, H. (2012). Let us eat cake? Historically reframing the problem of world hunger and its purported solutions. In C. Rosin, P. Stock, & H. Campbell (Eds.), *Food systems failure, the global food crisis and the future of agriculture* (pp. 30–45). London: Earthscan.

Campbell, H., & Fitzgerald, R. (2001). Follow the fear: A multi-sited approach to GM. *Rural Society, 11*(3), 211–224.

Campbell, H., Evans, D., & Murcott, A. (2017). Measurability, austerity and edibility: Introducing waste into food regime theory. *Journal of Rural Studies, 51*, 168–177.

Cardello, A. V. (2003). Consumer concerns and expectations about novel food processing technologies: Effects on product liking☆. *Appetite, 40*(3), 217–233.

Cox, D. N., & Evans, G. (2008). Construction and validation of a psychometric scale to measure consumers' fears of novel food technologies: The food technology neophobia scale. *Food Quality and Preference, 19*(8), 704–710.

Derry, T. K., & Williams, T. I. (1960). *A short history of technology from the earliest times to AD 1900*. Oxford: Clarendon press.

Dixon, J. (2009). From the imperial to the empty calorie: How nutrition relations underpin food regime transitions. *Agriculture and Human Values, 26*(4), 321–333.

FAO. (1996). *Rome declaration on world food security and world, food summit plan of action*. Rome: FAO.

Floros, J. D., Newsome, R., Fisher, W., Barbosa-Cánovas, G. V., Chen, H., Dunne, C. P., German, J. B., Hall, R. L., Heldman, D. R., Karwe, M. V., & Knabel, S. J. (2010). Feeding the world today and tomorrow: The importance of food science and technology. *Comprehensive Reviews in Food Science and Food Safety, 9*(5), 572–599.

Freidberg, S. (2009). *Fresh: A perishable history*. Cambridge, MA: Harvard University Press.

Friedmann, H. (1993). After Midas's feast: Alternative food regimes for the future. In P. Allen (Ed.), *Food for the future: Conditions and contradictions of sustainability* (pp. 213–233). New York: Wiley.

Friedmann, H. (2005). From colonialism to green capitalism: Social movements and emergence of food regimes. In *New directions in the sociology of global development*, ed. F. Buttel, and P. McMichael, 227–264. Amsterdam: Elsevier.

Friedmann, H. (2009). Discussion: Moving food regimes forward: Reflections on symposium essays. *Agriculture and Human Values, 26*(4), 335–344.

Friedmann, H., & McMichael, P. (1989). Agriculture and the state system: The rise and decline of national agricultures, 1870 to the present. *Sociologia Ruralis, 29*(2), 93–117.

Friedmann, H., & McNair, A. (2008). Whose rules rule? Contested projects to certify 'local production for distant consumers'. *Journal of Agrarian Change, 8*(2–3), 408–434.

Godfray, H. C. J., Beddington, J. R., Crute, I. R., Haddad, L., Lawrence, D., Muir, J. F., Pretty, J., Robinson, S., Thomas, S. M., & Toulmin, C. (2010). Food security: The challenge of feeding 9 billion people. *Science, 327*(5967), 812–818.

Goodman, D., & Goodman, M. (2009). Alternative food networks. In R. Kitchin & N. Thrift (Eds.), *International encyclopedia of human geography 3* (pp. 208–220). London: Elsevier.

Hajkowicz, S. (2015). *Global megatrends: Seven patterns of change shaping our future*. Clayton South: CSIRO Publishing.

Hawkes, C. (2009). Uneven dietary development: Linking the policies and processes of globalization with the nutrition transition, obesity and diet-related chronic diseases. *Globalization and Health, 2*(1), 4.

Holt Giménez, E., & Shattuck, A. (2011). Food crises, food regimes and food movements: Rumblings of reform or tides of transformation? *The Journal of Peasant Studies, 38*(1), 109–144.

Huang, H. T. (2000). Biology and biotechnology. In J. Needham (Ed.), *Science and civilization in China: Volume 6, biology and biological technology, Part 5, fermentations and food science*. Cambridge, UK: Cambridge University Press.

Jaeger, H., Knorr, D., Szabó, E., Hámori, J., & Bánáti, D. (2015). Impact of terminology on consumer acceptance of emerging technologies through the example of PEF technology. *Innovative Food Science & Emerging Technologies, 29*, 87–93.

Jermann, C., Koutchma, T., Margas, E., Leadley, C., & Ros-Polski, V. (2015). Mapping trends in novel and emerging food processing technologies around the world. *Innovative Food Science & Emerging Technologies, 31*, 14–27.

Jia, D. Y., Li, Y., Yao, K., & He, Q. (2005). Extraction of polyphenols from banana peel. *Journal of Sichuan University Engineering Science Edition, 37*(6), 52.

Kanatt, S. R., Chander, R., Radhakrishna, P., & Sharma, A. (2005). Potato peel extract a natural antioxidant for retarding lipid peroxidation in radiation processed lamb meat. *Journal of Agricultural and Food Chemistry, 53*(5), 1499–1504.

Karunasagar, I., & Karunasagar, I. (2016). Challenges of food security–need for interdisciplinary collaboration. *Procedia Food Science, 6*, 31–33.

Levidow, L. (2015). European transitions towards a corporate-environmental food regime: Agroecological incorporation or contestation? *Journal of Rural Studies, 40*, 76–89.

Liu, Y., & Lopez, R. A. (2016). The impact of social media conversations on consumer brand choices. *Marketing Letters, 27*(1), 1–13.

Masanet, E., Worrell, E., Graus, W., Galitsky, C. (2008). Energy efficiency improvement and cost saving opportunities for the fruit and vegetable processing industry. *An Energy Star Guide for Energy and Plant Managers*.

McMichael, P. (2005). Global development and the corporate food regime. *Research in Rural Sociology and Development, 11*, 269–303.

McMichael, P. (2009). A food regime analysis of the 'world food crisis'. *Agriculture and Human Values, 26*(4), 281–295.

Mirabella, N., Castellani, V., & Sala, S. (2014). Current options for the valorization of food manufacturing waste: A review. *Journal of Cleaner Production, 65*, 28–41.

Mohapatra, D., Mishra, S., & Sutar, N. (2010). Banana and its by-product utilization: An overview. *Journal of Scientific and Industrial Research, 69*(5), 323–329.

Oser, B. L. (1978). Benefit/risk: Whose? what? how much? *Food Technology (USA), 32*, 55–58.

Oyewo, O. A., Onyango, M. S., & Wolkersdorfer, C. (2016). Application of banana peels nanosorbent for the removal of radioactive minerals from real mine water. *Journal of Environmental Radioactivity, 164*, 369–376.

Pereira, R. N., & Vicente, A. A. (2010). Environmental impact of novel thermal and non-thermal technologies in food processing. *Food Research International, 43*(7), 1936–1943.

Pingali, P. L. (2012). Green revolution: Impacts, limits, and the path ahead. *Proceedings of the National Academy of Sciences, 109*(31), 12302–12308.

Popkin, B. (2003). The nutrition transition in the developing world. *Development Policy Review, 21*(5–6), 581–597.

Pritchard, B. (2009). Food regimes. In R. Kitchin & N. Thrift (Eds.), *International encyclopedia of human geography* (pp. 221–225). London: Elsevier.

Putnik, P., Bursać Kovačević, D., Režek Jambrak, A., Barba, F., Cravotto, G., Binello, A., Lorenzo, J., & Shpigelman, A. (2017). Innovative "green" and novel strategies for the extraction of bioactive added value compounds from citrus wastes—A review. *Molecules, 2*.

Qiu, L. P., Zhao, G. L., Wu, H., Jiang, L., Li, X. F., & Liu, J. J. (2010). Investigation of combined effects of independent variables on extraction of pectin from banana peel using response surface methodology. *Carbohydrate Polymers, 80*(2), 326–331.

Rendueles, E., Omer, M. K., Alvseike, O., Alonso-Calleja, C., Capita, R., & Prieto, M. (2011). Microbiological food safety assessment of high hydrostatic pressure processing: A review. *LWT-Food Science and Technology, 44*(5), 1251–1260.

Rosegrant, M. W., & Cline, S. A. (2003). Global food security: Challenges and policies. *Science, 302*(5652), 1917–1919.

Sage, C. (2013). The interconnected challenges for food security from a food regimes perspective: Energy, climate and malconsumption. *Journal of Rural Studies, 29*, 71–80.

Siegrist, M. (2003). Perception of gene technology, and food risks: Results of a survey in Switzerland. *Journal of Risk Research, 6*(1), 45–60.

Siegrist, M. (2008). Factors influencing public acceptance of innovative food technologies and products. *Trends in Food Science & Technology, 19*(11), 603–608.

Sims, R. E., Rogner, H. H., & Gregory, K. (2003). Carbon emission and mitigation cost comparisons between fossil fuel, nuclear and renewable energy resources for electricity generation. *Energy Policy, 31*(13), 1315–1326.

Slovic, P. (1987). Perception of risk. *Science, 236*(4799), 280–285.

Sparks, P., & Shepherd, R. (1994). Public perceptions of food-related hazards: Individual and social dimensions. *Food Quality and Preference, 5*(3), 185–194.

Teuteberg, H. J. (1995). History of cooling and freezing techniques and their impact on nutrition in twentieth century Germany. In A. P. den Hartog (Ed.), *Food technology, science and marketing. European diet in the twentieth century*. East Lothian: Tuckwell Press.

Thorne, S. (1986). *The history of food preservation*. Cumbria: Parthenon Publishing.

Truninger, M. (2013). The historical development of industrial and domestic food technologies. In A. Murcott, W. Belasco, & P. Jackson (Eds.), *The handbook of food research* (pp. 82–108). London: Bloomsbury.

Van Boekel, M., Fogliano, V., Pellegrini, N., Stanton, C., Scholz, G., Lalljie, S., Somoza, V., Knorr, D., Jasti, P. R., & Eisenbrand, G. (2010). A review on the beneficial aspects of food processing. *Molecular Nutrition & Food Research, 54*(9), 1215–1247.

van der Goot, A. J., Pelgrom, P. J., Berghout, J. A., Geerts, M. E., Jankowiak, L., Hardt, N. A., Keijer, J., Schutyser, M. A., Nikiforidis, C. V., & Boom, R. M. (2016). Concepts for further sustainable production of foods. *Journal of Food Engineering, 168*, 42–51.

Wang, C. Y., Huang, H. W., Hsu, C. P., & Yang, B. B. (2016). Recent advances in food processing using high hydrostatic pressure technology. *Critical Reviews in Food Science and Nutrition, 56*(4), 527–540.

Weaver, C. M., Dwyer, J., Fulgoni, V. L., King, J. C., Leveille, G. A., MacDonald, R. S., Ordovas, J., & Schnakenberg, D. (2014). Processed foods: Contributions to nutrition. *The American Journal of Clinical Nutrition, 99*(6), 1525–1542.

Williams, S. N., & Nestle, M. (2015). 'Big food': Taking a critical perspective on a global public health problem. *Critical Public Health, 25*(3), 245–247.

Wobiwo, F.A., Emaga, T.H., Fokou, E., Boda, M., Gillet, S., Deleu, M., Richel, A., Gerin, P.A. (2017). Comparative biochemical methane potential of some varieties of residual banana biomass and renewable energy potential. *Biomass Conversion and Biorefinery, 7*(2), 167–177.

Zachmann, K. (2011). Atoms for peace and radiation for safety: How to build trust in irradiate foods in cold war Europe and beyond. *History and Technology: An International Journal, 27*(1), 65–90.

Zink, D. L. (1997). The impact of consumer demands and trends on food processing. *Emerging Infectious Diseases, 3*(4), 467–469.

Chapter 2
Overview on the Food Industry and Its Advancement

Mehrdad Niakousari, Sara Hedayati, Zahra Tahsiri, and Hamide Mirzaee

2.1 Introduction

Interest has grown in the modern world in foods that are safe, nutraceutical, composed of more natural colors and flavors, and produced by environmentally friendly methods. The key to the continuous growth of the food industry is innovations in food processing technologies that target the evolving consumer interests. Food processing is moving towards environmental sustainability by means of employing novel technologies that reduce water and energy consumption (Knoerzer et al. 2015). Such novel methods may also have additional advantages such as preserving nutritional value and enhancing food quality (Barbosa-Cánovas et al. 2011). Since, nowadays consumers are more health cautious and focused on what they eat and how it is produced compared to a few decades ago, innovative approaches also allow producers to meet the mounting demands of the competitive global market. The novel techniques currently employed in the food industry are classified into thermal and non-thermal processing. The capabilities, applications, advantages and limitations of these innovative techniques have been concisely discussed in detail in this chapter.

2.2 Non-thermal Food Processing

Non-thermal processing refers to techniques that are effective at sublethal or ambient temperatures. They have the advantage of saving energy while destroying pathogens. In addition, they can increase the shelf life and preserve the nutrients to greater

M. Niakousari (✉) · S. Hedayati · Z. Tahsiri · H. Mirzaee
Department of Food Science and Technology, School of Agriculture, Shiraz University, Shiraz, Iran
e-mail: niakosar@shirazu.ac.ir; sara.hedayati@mail.um.ac.ir; hmirzaee@ut.ac.ir

© Springer Nature Switzerland AG 2019
C. Piatti et al. (eds.), *Food Tech Transitions*,
https://doi.org/10.1007/978-3-030-21059-5_2

extend in comparison to conventional approaches (Cullen et al. 2012; de Toledo Guimarães et al. 2018). These technologies include pulsed electric fields (PEF), supercritical CO_2, high pressure processing (HPP), radiation, ozone processing, etc.

2.2.1 Pulsed Electric Fields (PEF)

PEF involves the extremely brief (μs to ms) use of electric fields with high voltage (usually 20–80 kV/cm). This technique has to an extent replaced conventional methods for the inactivation of microorganisms and enzymes with the aim of preserving food in an environmentally friendly way that avoids the effects of heating. The PEF method inactivates microorganisms owing to the force of the external electric field, which produces single or multiple pores in the microbial cell membranes and destabilizes the microorganisms (Grahl & Märkl 1996). In continuation, cell membrane functions are interrupted and intracellular contents pour out of the cells, effectively destroying the microorganism (De Silva et al. 2018). PEF was initially patented almost 30 years ago, but was first used for large-scale commercial purposes around 15 years ago. In this time, various highly invested groups in Europe and the United States have worked hard to incorporate PEF into the food industry, and researchers around the world have published a large amount of papers on this novel technology. However, despite these widespread efforts, the food industry has only commercially used PEF to a limited extent. A minimum of six different manufacturers have introduced systems of PEF technology in the past two decades, with an average power rating of between 3 and 600 kW and pasteurization capacities of between 3 and 10,000 L/h. Some of these manufacturers work exclusively on PEF, including the Elea company in Germany, the Scandinova company in Sweden, and the Diversified Technologies Incorporation (DTI) in the United States (Kempkes et al. 2016).

2.2.2 High Pressure Processing (HPP)

Another non-thermal (cold pasteurization) food processing technique is HPP, which preserves foods without the need for additives. HPP has proven itself as a method for producing food products that are safe (pathogen-free) and stable. In addition, this technology improves the quality characteristics of food products such as color and flavor. Although the principles upon which HPP acts to inactive microbes have been known for over a century, the HPP technology has been employed in wide scale only in the past two decades (Muntean et al. 2016). By this technology, pressures of up to 600 MPa are applied to inactivate vegetative bacteria, molds and yeast at ambient temperatures. Moreover, spores can be inactivated at high temperatures in a process known as high pressure thermal processing (HPTP) (Jermann et al. 2015). Food preservation is the prime application of HPP in the food industry. Food products are usually spoiled by enzymes, via the catalysis of biochemical processes, and microorganisms. HPP combats this problem by inactivating most of such enzymes and microorganisms. On the other hand, HPP has only a minimal effect on low molecular weight molecules. Hence, useful substances such as pigments, vitamins and flavor compounds remain largely undamaged relative to thermal processing. However, an irreversible change is made to some compounds by means of HPP. Macromolecules such as proteins and carbohydrates are considerably affected by HPP. Proteins are not resistant to high pressures and may be denatured during HPP while carbohydrates can be gelatinized by applying increased pressure instead of elevated temperature (Muntean et al. 2016). HPP is already used at an industrial scale for processing of different types of fish and meat products. However, the high pressure levels required for microbial inactivation may affect texture or aroma of these products and decrease their quality characteristics. This effect has been attributed to the protein denaturation or lipid oxidation under HP conditions. Hence a choice between microbiological safety and sensory attributes is a challenge in the commercial application of HPP. Nevertheless, the careful adjustment of processing conditions such as temperature, pressure, kinetics, the formulation of the product and packaging may counteract the adverse effects of HPP (Duranton et al. 2014).

2.2.3 Irradiation

Irradiation, the application of ionizing radiation, is a flexible and effective non-thermal antimicrobial process used in food processing (Smith and Pillai 2004). High-energy photons such as X-rays or gamma rays, energize the electrons in the atoms of foods. These electrons may leave the atom which is known as ionization or may increase the energy of electrons which is known as excitation. These processes produce free radicals which are very reactive, due to their unpaired electrons which pair up with outer shell electrons of atoms. Water makes up the bulk of mass in most of food materials, thus most of the energy absorption from irradiation changes water molecules into hydroxyl and hydrogen radicals (Diehl 1999). The interaction between organic molecules within food materials and free radicals is the chief mode of action in irradiation. When the degree of free water is limited (e.g. in frozen or dried products) less free radicals are formed per unit of energy applied. Moreover, the mobility of radicals is reduced. Hence, higher doses of irradiation are necessary for the microbial safety in such products (Thayer and Boyd 1995). Food irradiation has been endorsed by international organizations such as Food and Agriculture Organization of the United Nations (FAO), US General Accounting Office (GAO 2000), the American Dietetic Association (ADA 2000) and the UN World Health Organization (WHO 1999) as a safe and effective food-processing treatment. Moreover, it has been addressed in international standards such as Codex Alimentarius Standard and in standards of the International Plant Protection Convention (IPPC) (Carreño 2017; Niemira and Gao 2012). Many food products are responsible for the occurrence of enterohemorrhagic *E. coli* infection, and hemolytic uremic syndrome and salmonellosis (FDA 2009), therefore processes which ensure their microbial safety are required. Although traditional heat treatments such as pasteurization ensures the microbial safety of food products, they result in the loss of textural attributes, vitamins, color, essential oils, flavor and aroma which play the major role in acceptability of food products. On the contrary, irradiation saves the organoleptic and nutritional properties of food products while ensuring their safety. Moreover, it is considered as the most applicable post-packaging treatment. The application of ionizing radiation for food preservation goes back to more than a century. However, its first industrial use, was around the 1950s and was proliferated after the 1960s in the US. Ionizing radiation was first used for the production of sterile meat products in order to substitute frozen and canned military rations (Blackburn 2017; Ruan et al. 2001). Irradiation has been shown to be effective for improving the microbial safety of fresh, chilled or frozen meat and poultry, fruits, vegetables, seasonings, spices, herbs, animal feed and pet food (Mittendorfer 2016). Other applications of irradiation include decomposition of toxins such as ochratoxin A in wheat flour, grape juice and wine (Calado et al. 2018) and aflatoxin B1 in soybean (Zhang et al. 2018), decomposition of antibiotics (Amoxicillin, Doxycycline, and Ciprofloxacin) in milk, chicken meat and eggs, insect disinfestation in wheat and wheat flour, sprout inhibition in potatoes, control *Trichinella spiralis* in pork and delay ripening in fruits (Mittendorfer 2016).

2.2.4   Ozone Processing

Ozone is a natural gas which is found in the atmosphere, however it can also be formed synthetically. It contains three atoms of oxygen (O_3) which has an isosceles triangle form (Muthukumarappan 2011). Ozone is relatively unstable under normal temperature and pressure conditions and is the strongest antimicrobial agent for food contamination (Mahapatra et al. 2005; Muthukumarappan et al. 2000, 2002). The third oxygen atom in ozone molecule is electrophilic and has a small free radical electrical charge. To balance itself electrically, the third oxygen atom seeks for materials with unbalanced opposite charge. Normal healthy cells have a strong enzyme system and a balanced electrical charge thus cannot interact with ozone or its derivatives. While the pathogens or cells stressed by microorganisms are electropositive and carry such opposite charge and attract ozone or its by-products (Muthukumarappan 2011). Ozone Processing is a powerful disinfectant against various microorganisms such as bacteria, viruses, fungi, protozoa, fungal and bacterial spores (Khadre et al. 2001). Ozone inactivates microorganisms by oxidization. The disinfection ability of ozone in foods can be influenced by extrinsic and intrinsic factors, such as pH of food, relative humidity, temperature and presence of organic compounds in food matric (Kim et al. 1999). The ozone residues decompose to non-toxic products such as oxygen. Therefore, it is considered as an eco-friendly antimicrobial agent for industrial applications (Kim et al. 1999). The strong antimicrobial properties of ozone are mainly due to its high oxidizing potential and diffusion through cell membranes (Hunt and Mariñas 1999). The biocidal effects of ozone are 1.5 times more than chlorine and are useful for inactivation of a wide range of microorganisms (Xu 1999). Ozonation has been approved as an antimicrobial agent in food products by FDA and is recognized as Generally Recognized as Safe (GRAS) substance for food industry applications. Consequently, it has received much attention in processing of different food products. The industrial applications of ozone in the food processing include washing, recycling of poultry wash water, vegetable processing, seafood sterilization, sanitation of equipment and surfaces in beverage manufacturing industry, and the treatment of bottled drinking water (Doona 2010; Patil and Bourke 2012). Presently, there are more than 3000 ozone-based water treatment installations in the world and more than 300 potable water treatment plants in the US (Patil and Bourke 2012; Rice et al. 2000). Other applications of ozone have been listed in Table 2.1.

Table 2.1 Applications of ozone treatment in food processing

Product	Target	Result	Reference
Corn	Degradation of aflatoxin B1 (AFB1) in corn	Ozone treatment is a safe, effective and fast method for the degradation of AFB1 and the degradation rate increased by the concentration of ozone and treatment time	Luo et al. (2014)
Potato starch	Starch modification	Ozone is effective in the modification of potato starch structure and functional properties due to its oxidizing properties	Castanha et al. (2017)
Carrot	Evaluation of ozone type (gas or dissolved in water) on the quality of carrots	Ozone can be used for carrots processing in the form of gas and dissolved in water. However, ozone gas was more effective for shelf life extension of carrots	de Souza et al. (2018)
Grains	Studying the impact of ozone treatment on mycotoxins and physicochemical properties of wheat, rice and maize	Ozone improves the functional properties of grains while ensuring their safety	Zhu (2018)
Cantaloupe melon juice	Assessment of juice quality pasteurized by gaseous ozone	Ozone is a useful treatment in low contaminated juice but has negative effects on its physicochemical and nutritional properties	Fundo et al. (2018)
Water and waste milk	Removal of antibiotics	Ozonation decomposed the antibiotics and their decomposition in milk was more effective than in water	Alsager et al. (2018)
Wheat seed	Reduction of insects and microorganisms	Ozone has the potential to reduce insects and contaminating microorganisms in grains with slight or no effect on their quality	Granella et al. (2018)

Besides the advantages of ozone treatments, it has some shortcomings which limit its applications in food industry. Some of these disadvantages have been noted by Muthukumarappan et al. (2000) as listed below:

- Ozone is a poisonous gas which can cause irritation to the throat and nose, loss of vision, pain in the chest, headache, cough, drying of the throat, increase in heartbeat, edema of lungs or even death if inhaled in high extent.
- Doses of ozone that are large enough for decontamination of food products may negatively affect their color, odor or nutritional quality.
- The presence of organic matter increases the ozone consumption and makes it difficult to predict the supply of ozone required for elimination of microbial population in food products.
- In many fruits and vegetables, aqueous ozone should be used for decontamination which require a large amount of water and techniques to overcome the gas/liquid interface.
- Ozone has a short shelf-life; therefore, the prepared aqueous ozone should be used immediately.

To overcome these drawbacks and achieve the target microbial inactivation the mechanism of the interactions between ozone and food components should be studied precisely and the process parameters such as temperature, ozone concentration, etc. should be optimized.

2.2.5   Supercritical CO₂

Supercritical is a state of matter higher than its critical temperature and critical pressure (Chung et al. 2013). Supercritical fluids possess high solvation power due to their high density or liquid- like properties, and high diffusivity due to low viscosity values or gas- like properties (Zabot et al. 2014). Carbon dioxide (CO_2), methanol, methane, ethanol, ethane, acetone, and propane are different supercritical fluids used in industry (Ilgaz et al. 2018). Among them, CO_2 is considered to be a low cost, inert, easily available, odorless, tasteless, environmentally friendly, non-flammable, GRAS solvent and has low critical pressure (73 bar) and low critical temperature (304 K) (Chung et al. 2013; Rahman et al. 2018; Soares and Coelho 2012). Supercritical CO_2 is a desirable, green, non-thermal technology for food processing industries by maintaining nutritional, physicochemical and organoleptic properties of fresh food products (Amaral et al. 2018). Nowadays, the supercritical fluid technology has many applications in the food industry such as decaffeination, sterilization and extraction of bioactive compounds and about more than 150 plants profit from this technology worldwide (Chung et al. 2013; del Valle 2015; Knez et al. 2014). Coffee and tea decaffeination and hop flavor extraction are the common industrial applications of supercritical CO_2 (del Valle 2015).

Based on literature, Supercritical CO_2 is a promising technique for extraction (Derrien et al. 2018) pasteurization (Di Giacomo et al. 2016), sterilization (Di Giacomo et al. 2016; Perrut 2012), drying (Knez et al. 2014; Oualid et al. 2018) and enzyme inactivation (Omar ct al. 2018) processes. Supercritical carbon dioxide systems can operate in batch, semi- continuous and continuous modes (Omar et al. 2018). In supercritical fluid technology, pressure vessel, as the process chamber, is the most important part of system (Soares and Coelho 2012).

Supercritical extraction led to a fast, high selective, non-thermal process with no solvent residue in product in comparison to conventional procedures, solvent extraction method (Soares and Coelho 2012; Valadez-Carmona et al. 2018). Investigations on extraction yield of supercritical carbon dioxide demonstrated that there are any differences between laboratory scale and industrial scales, but lower extraction in case of kinetic industrial scales is not avoidable because of the greater size of solid particles (García-Risco et al. 2011).

The drawback of using supercritical CO_2 in extraction processing is its nonpolar characteristic that limits polar compounds extraction (Zabot et al. 2014). In order to overcome this limitation, use of polar co- solvent such as methanol, ethanol, hexane, acetone, chloroform and water is a common procedure (Rawson et al. 2012; Tello et al. 2011). Among this polar co- solvents, ethanol is the most common because of its low toxicity and easily removable characteristics.

Briefly, antimicrobial effect of supercritical CO_2 may be due to different mechanisms. Key enzymes inactivation, intercellular pH decreasing and effect on cell membrane are the most important mechanisms (Rawson et al. 2012).

In spite of application of this technology in industry since 1950s, and its advantages, there is also some resistance to upgrade conventional technologies to supercritical carbon dioxide systems as an emerging technology because of expensive initial installation cost and risk of equipment explosion due to high pressure utilization (del Valle 2015; Rahman et al. 2018; Tello et al. 2011; Zhang et al. 2014). Optimization of process variables and bed geometry are of the most important factors for decreasing operation costs by increasing the yield and decreasing processing time (Tello et al. 2011; Zabot et al. 2014). Moreover, elimination of solvent removal step and recovery of carbon dioxide are some of the factors would reduce the operation cost (Subramaniam 2017). Table 2.2 describes some of supercritical CO_2 applications in food industry.

2.3 Thermal Processing

Novel thermal processing such as microwave, ohmic heating (OH) and radio frequency (RF) heating have been considered as alternatives to traditional heat treatments in recent years.

Table 2.2 Some of supercritical CO_2 applications in food industry

Process	Material	Purpose	Treatment conditions	Results	Reference
Ethanol-modified supercritical CO_2	Camelina sativa seed	Extraction of astaxanthin	Pressure: 41.6 MPa, temperature: 36.6 °C, ethanol concentration: 42.0%	The extracted oil contained higher astaxanthin level and antioxidant activity	Xie et al. (2019)
Supercritical CO_2	Bovine heart	Lipid extraction	Pressure: 30 and 40 MPa, temperature: 40 °C	Unsaturated fatty acids were higher (53.09 g/100 g fatty acids) than the control (45.62 g/100 g fatty acids) and hexane-extracted samples. Moreover, microbes were inactivated and low-fat protein rich meat was produced	Rahman et al. (2018)
Supercritical CO_2	Carrot juice	Preservation	Pressure: 25 MPa, temperature: 313 K, juice/ CO_2 ratio (w/w): 0.33	Peroxidase was inactivated completely while inactivation rate in case of pectinesterase: was above 70%	Di Giacomo et al. (2016)
Supercritical CO_2 and high power ultrasound	Coriander	Microbial inactivation and drying	Power: 40 W, pressure: 10 MPa, temperature: 40 °C	Dehydrated products presented satisfying microbial content reduction	Michelino et al. (2018)
Supercritical CO_2	Fresh palm fruits	Waterless sterilization	Pressure: 8–40 MPa, temperature: ≤60 °C	Enzyme and lipophilic microorganisms were inactivated	Omar et al. (2018)

2.3.1 Microwave

Microwaves are electromagnetic waves with frequencies between 1 and 30 GHz. Microwave heating is amongst the methods of thermally processing of food products (Meda et al. 2017). With increased standards of living and greater incomes, consumers have demanded novel food processing methods in recent years. Microwave heating has gained popularity due to being a revolutionary technology that preserves the nutritional value of food products (Kalla 2017; Meda et al. 2017). It is well-recognized in terms of its operational safety and large capacity for the retention of nutrients, meaning that heat sensitive nutrients including carotenoids, dietary antioxidants vitamins B and C, and phenols are sparingly lost during heating (Kalla 2017). Various industries have successfully employed microwave heating for food processing. In comparison to conventional methods, the pasteurization and sterilization of food products via microwave heating is claimed to destroy pathogens more efficiently, with significantly less damage to product quality and significantly faster processing (Zhu et al. 2018b). Microwave heating has various applications in food processing, including baking, blanching and cooking. When compared with conventional methods, microwave heating retains a greater amount of nutritional value, quality, taste and flavor, while also preserving food products more effectively. This technology has also dramatically decreased the amount of energy used for food drying (Kalla 2017).

2.3.2 Ohmic Heating (OH)

OH is an alternative thermal processing method for pasteurization, sterilization and cooking of food products that is considered as a high temperature short time process. During OH an alternating current (typically 50 Hz to 100 kHz) passes through the food material and generates a uniform temperature profile inside food due to its electrical resistance. Therefore, the drawbacks such as fouling, deterioration of product quality due to overheating, and having difficulty in heating viscous foods or products with solid fractions which are observed in conventional heat treatments would not occur in OH. OH is effective in saving energy, nutritional and sensory properties of foods and is a more efficient technique in inactivation of spoilage and pathogenic microorganisms (Ruan et al. 2001). One of the potential applications of OH is in the peeling of fruits and vegetables which eliminates the need of lye (Ramaswamy et al. 2005). Thus, it can be considered as an environmentally friendly peeling technique. These features make OH a suitable choice for susceptible foods, liquids containing particles, slurries, and highly viscous materials. OH technology has been known since the nineteenth century and the very first application of OH in food industry was for the pasteurization of milk (Anderson and Finkelstein 1919). But it was abandoned due to high processing costs and lack of inert materials for the electrodes. In the 1980s OH was revived due to the accessibility of improved electrode materials. However, the initial operational costs and lack of knowledge on validation procedures prevent its extensive application in food industry. Furthermore, dependency on electrical resistance may be a disadvantage in heating products which are not ionically loaded (e.g. distilled water, oil and products with high fat content) (Leadley 2008). Currently, OH has been used for a variety of applications as observed in Table 2.3. It is estimated that around 100 commercial plants in the USA, Mexico, Europe and Japan are using OH technique (Leadley 2008). OH is used in an industrial scale in Japan, England, US and Italy for different purposes including pasteurization of liquid egg, meat and vegetable products, to process ready-to-eat sauces, fruit juices, fruit nectars, fruit purée, fruit slices in syrups, tomato paste, soups, meat products, stews, sauces and treatment of heat sensitive liquid materials (Leadley 2008; Mans and Swientek 1993).

Table 2.3 Applications of ohmic heating (OH) in food processing

Product	Target	Result	Reference
Jerusalem artichoke	Extraction of inulin	OH resulted in higher inulin extraction yield compared to conventional heating method	Termrittikul et al. (2018)
Orange juice waste	Extraction of pectin	OH increased the pectin extraction yield more than the conventional heating	Saberian et al. (2017)
Colored potato	Extraction of phytochemicals	OH enhanced extraction of phytochemicals, reduced treatment time and required lower power consumption with no organic solvents (green extraction)	Pereira et al. (2016)
Fermented red pepper paste	Pasteurization	The quality of the samples pasteurized by OH was higher than conventionally heated samples and a 99.7% reduction was observed in Bacillus strains by OH while this value was 81.9% when conventional heating was used	Cho et al. (2016)
Grapefruit and blood orange juices	Pasteurization	OH preserved the carotenoid profile of citrus fruits	Achir et al. (2016)
Blueberry pulp	Heat treatment	The anthocyanin degradation level in samples treated with low voltage OH was lower or similar to conventional heating. However, degradation was increased at higher voltages	Sarkis et al. (2013)
Pomegranate juice	Concentration	OH is an energy efficient system and concentration time was shortened about 56% by OH	Icier et al. (2017)
Sour cherry juice	Concentration	Evaporation was performed successfully and the process time was shortened	Sabanci and Icier (2017)
Shrimps	Cooking	The cooking times of shrimps was reduced by ≈50% in OH and the treatment was more uniform with less color differences compared to conventional cooking	Lascorz et al. (2016)
Rice	Cooking	OH saves more than 70% of energy in comparison with a commercial rice cooker. Moreover, no fouling of rice was observed on the container after cooking	Kanjanapongkul (2017)
Carrot, golden carrot and red beet	Cooking (texture softening)	OH resulted in greater softening rates and can be applied for modification of vegetables texture	Farahnaky et al. (2012)
Pumpkin	Blanching	OH resulted in a faster inactivation of peroxidase, however similar changes were observed in pumpkin color	Gomes et al. (2018)
Tomato	Peeling	OH is an effective treatment for tomato peeling and it is possible to decrease the lye concentration by using ohmic peeling	Wongsa-Ngasri & Sastry (2015)

2.3.3 Infrared (IR) Heating

IR radiation is an electromagnetic radiation and alternative technology for thermal treatment of food materials (Pawar and Pratape 2017). IR radiation possesses many advantages including equipment simplicity, high speed, low capital cost, uniform heating, low quality deterioration, radiation energy transmission without heating the sample's surrounding air (Adak et al. 2017; Gili et al. 2017; Irakli et al. 2018; Kettler et al. 2017; Van Bockstal et al. 2017; Yılmaz 2016).

Based on IR radiation energy source temperature, three different IR radiation categories are the far-infrared (FIR) (3–1000 µm), middle-infrared (MIR) (1.4–3 µm) and near-infrared (NIR) (0.78–1.4 µm). As the wavelength increases, the penetration depth increases too (Riadh et al. 2015). Far-IR radiation is effective by damaging RNA, DNA, cell proteins and ribosomes to ensure food safety and is applied in pasteurization and sterilization of foods (Hu et al. 2017).

Product energy absorbance depends on the irradiation wavelength and the surface characteristics (Adak et al. 2017). Absorption characteristics of product surface determine the process efficiency (Pawar and Pratape 2017). When the material surface receives IR radiation energy, it transfers to heat and generates into the material with conduction (Gili et al. 2017).

Some of the potential IR applications in food industry are its use for drying, roasting, frying, baking, pasteurization, peeling and blanching (Kettler et al. 2017; Yalcin and Basman 2015). Table 2.4 summarizes some of the application and process conditions of IR heating. Among these, high-quality IR drying of fruit, nuts, and grains, lonely or in combination with other drying methods is its most popular application (Ding et al. 2015). The most important limitation of IR radiation is its low penetration depth that depends on energy wavelength and product characteristics. In order to effective heating of whole food materials, combining IR with heating technologies such as microwave, hot air and other common techniques has been suggested (Adak et al. 2017; Riadh et al. 2015).

2.3.4 Radio Frequency (RF) Heating

RF as a nonionizing electromagnetic radiation is a novel thermal technology in the food industry (Jiang et al. 2018; Pereira and Vicente 2010). Frequencies between 3 and 300 MHz refers to RF but only 40.68, 27.12 and 13.56 MHz are used for medical, scientific and industrial applications, respectively (Huang et al. 2016; Ozturk et al. 2016).

As for microwave heating, the mechanism of temperature rising in RF radiated substance is based on dipole rotation and ionic depolarization and as a consequence, volumetric heating (Kirmaci & Singh 2012; Pereira and Vicente 2010). Volumetric heating reduces heating time, quality deterioration and overheating in comparison to conventional thermal methods and makes RF heating a promising method for

Table 2.4 Infrared (IR) heating applications in food industry

Process	Purpose	Material	Treatment conditions	Results	Reference
Mid-IR radiation in combination with freeze-drying	Drying	Mushroom	Freeze drying time: 4 h Mid-infrared drying after the freeze drying stage at 60 °C until the <12% (w/w) moisture content was reached	IR is a time saving drying process which retains the aroma of products	Wang et al. (2015)
IR radiation heating	Drying	Strawberry	Power: 200 W Temperature: 100 °C Velocity: 1.5 m.s⁻¹	IR drying resulted in retention of nutrients and bioactive compounds and offered good consumer acceptability	Adak et al. (2017)
Hot air and sequential IR	Decontamination and drying	Almond	Hot air and IR drying temperature: 70 °C Tempering: 70 °C	Drying time was decreased in comparison with hot air drying alone Desirable almond oil peroxide value was observed and *Enterococcus faecium* population was reduced	Venkitasamy et al. (2018)

IR radiation in combination with abrasive unit	Dry-peeling system	Hazelnut	Radiation time: 3 min
			Radiation power: 1600 W
			Hazelnuts moisture content: 4% (w/w)
			Abrasive unit:
			  Clearance: 4 mm
			  Rotor speed: 200 rpm
			  Abrasive path length: 36 cm
				Process efficiency was: 81.6% and breakage of kernel was 0.87%	Eskandari et al. (2018)

Middle IR radiation	Stabilization	Rice bran	Power: 700 W Time: 7.0 min	Total tocopherol and Y-oryzanol contents of treated samples was higher than crude ones Rice bran shelf life was extended to 3 months	Yılmaz (2016)
IR radiation	Stabilization	Wheat germ	Radiation intensity: 4800 W/m² Time: 3 min Emitter-sample distance: 0.2 m	Lipase was inactivated, moisture content was reduced and shelf life was extended from 15 to 90 days at room temperature	Gili et al. (2017)

IR radiation	Shelf life improvement	Rough and brown rice	Radiation intensity: 4685 W/m² Temperature: 60 °C Tempering: 4 h	Lipase was inactivated and shelf life of brown rice at 35 ± 1 °C and 65 ± 3% relative humidity was extended for 10 months Milling quality was better than ambient and hot air dried samples	Ding et al. (2015)
IR radiation	Final cooking	Meatball	Time: 12 min Application distance: 10.5 cm	Final cooking by IR radiation improved color and texture of ohmically pre-cooked meatballs	Turp et al. (2016)
IR radiation	Roasting	Flaxseed	Radiation intensity: 1150 W/m²	Hydrogen cyanide (HCN) content of flaxseed was reduced 59%. Tocopherol content of flaxseed oil was increased and higher peroxide value was observed after 6 months of storage	Tuncel et al. (2017)

heating of the semi-solid and solid foods which have low thermal conductivities (Choi et al. 2018; Kim et al. 2012; Ozturk et al. 2016, 2018).

Advantages of RF heating relative to microwave and IR radiation is higher wavelength and penetration depth, which makes it suitable for large bulk and post-package food processing (Jiang et al. 2018; Liao et al. 2018; Zhu et al. 2018a). Due to these unique properties, RF heating is a desired technique for thawing (Erdogdu et al. 2017), drying (Wang et al. 2014), cooking (Schlisselberg et al. 2013), disinfestation (Wang et al. 2010; Zhou et al. 2015), pasteurization (Zheng et al. 2016; Zheng et al. 2017), roasting (Liao et al. 2018), enzyme inactivation (Ling et al. 2018), and tempering (Palazoğlu and Miran 2018), in the food processing industry. Some of the RF applications in food industry is given in Table 2.5.

Application of RF heating in food industry goes back to more than 60 years ago (Rincon et al. 2015). Industrial RF heating systems could be in continuous and batch modes, although continuous systems are more desirable in industrial scale (Erdogdu et al. 2017). Uneven temperature distribution and non- uniform heating is the most important drawback of RF heating prevents its commercialization (Alfaifi et al. 2016). The temperature non- uniformity is related to various factors such as physical, thermal and dielectric properties of food, food and packaging geometries, geometrical configuration of the sample, electrode shape, top electrode voltage, distance between the electrodes, distance between the top electrode and the sample (Erdogdu et al. 2017; Kim et al. 2012; Tiwari et al. 2011).

2.4   Standpoint of Food Experts About Commercialization of Novel Technologies

The acceptance of different food products manufactured by novel technologies is a critical issue in food industries and several researches have been performed to investigate the perception of people about such technologies (Behrens et al. 2009; Evans and Cox 2006). In a study by Jermann et al. (2015) the attitude of people about the potential of novel food processing technologies to be commercialized in future was investigated. To perform this study, two independent groups were designed. One group was based in North America (Survey 1) and the other was located in Europe (Survey 2). Food professionals from universities, industry and government were selected to respond to the surveys in order to identify the currently applied and up-coming food processing technologies, the related regulations, limitations, and factors pertaining to their commercialization. The results of both surveys revealed that microwave (88%), UV (84%) and HPP (80%) were the main currently applied and up-coming (in the next 5 years) food processing technologies in North America. In Europe however, PEF replaced UV in third place. The main motivators behind commercialization were the aims of producing products with higher quality (94%), greater safety (92%) and longer storage life (91%). The main technologies that were identified for current use and use in the next 10 years were HPP and microwave.

Table 2.5 Radio frequency (RF) applications in food industry

Process	Material	Purpose	Treatment conditions	Results	Reference
Hot air-assisted RF heating	Rice bran	Enzyme inactivation	Electrode gap: 10 cm	Residual enzymes activities was 19.2% for lipase and 5.5% for lipoxygenase.	Ling et al. (2018)
			Temperature: 100 °C		
			Hot air treatment time: 15 min		
Hot air-assisted RF heating	Cashew nut kernels	Roasting	Temperature: 120–130 °C	Moisture content of kernels was reduced from 6.2 g/100 g (d. b.) to 1.5 g/100 g within 30 min. Peroxide value and acid value were lower compared with hot air roasted samples (at 140 °C for 30 min)	Liao et al. (2018)
RF heating in combination with antimicrobial agents	Ground beef homogenate	Pasteurization	Temperature: 50 °C	E. coli was destroyed (5 log (CFU/ML))	Nagaraj et al. (2016)
Continues RF heating	Tuna fish	Thawing	Frequency: 27.12 MHz	Uniform and volumetric thawing was achieved	Erdogdu et al. (2017)
Continues RF heating	Lean beef	Tempering	Frequency: 27.12 MHz	Rapid and uniform tempering of beef was obtained	Palazoğlu and Miran (2018)
			Power: 2 kW		
Hot air assisted RF heating	Macadamia nuts	Drying	Frequency: 27.12 MHz	Drying was performed uniformly	Wang et al. (2014)

Moreover, geological differences were recognized with Europe having a greater focus on PEF while North America being more centered on UV and radiation technologies. In 10 years' time, Europe anticipated cold plasma and PEF to obtain greater importance while North American professionals thought HPP, microwave and UV will maintain their status as being most important (Jermann et al. 2015).

HPP is the most commercially used novel technology in the food industry and has been applied in an industrial scale in Japan since 1990s. HPP has been reported to be applied in vegetables, fruits, seafood and meat products in order to improve consumer acceptance, inactivation of bacteria, viruses and some of enzymes without affecting nutritional and organoleptic properties of food products. Moreover, experts have stated that HPP has potential to replace conventional heat treatments such as pasteurization (Alegbeleye et al. 2018). It has been documented that PEF technology has been used for commercial applications in different food processing sectors (Deeth and Lewis 2017). For example, the inactivation of vegetative microbial cells (up to 5–6 logs) has been reported for PEF which is identical to ultra-high temperature (UHT) sterilized milk in a very short time (microseconds) (Deeth and Lewis 2017).

2.5 Industrial Applications of Novel Technologies

Based on the unique properties and advantages of novel technologies, in recent years these processes have been applied in a commercial scale in food industry. Table 2.6 gives an overview about the products which are processed by novel technologies in food industry.

2.6 Conclusions

There have been substantial advancements in recent years in the progress of processes with commercial promise in food industry. The application of such treatments led to production of high-quality products, while reducing energy consumption and processing costs. Moreover, most of these technologies are clean and ecofriendly and have less environmental effect than the traditional ones due to replacement of non-renewable resources of energy by renewable ones, reduction of waste water and gas emission. It has also been proven that the application of emerging thermal and non-thermal treatments has the potential to produce safe and healthy food products. However, limitations, such as high investment costs, difficulty in adjustment and control of process operation variables and absence of regulatory approval have retarded a wider exploitation of novel technologies at an industrial scale. To take the advantages of such technologies in a commercial scale and expand their applications, issues associated with consumers' perception and optimization of processing conditions should be considered. To sum-up, some of the novel technologies are already used in food industry and it is expected that some other techniques find more extensive implementation in the food industry within the next few years and replacing or supplementing conventional processing due to being more efficient in preserving the nutritional characteristic of food products, their economic and technical advantages as well as being more environmental friendly.

Table 2.6 Industrial applications of novel technologies

Process	Product	Objectives	Food manufacturers	Equipment supplier
High pressure processing	Juice and beverages, smoothies, sliced and whole pieces of cured ham, chicken or Turkey cuts, ready-to-eat seafoods products, clams, oysters, mussels, crabs, lobsters, cod, shrimp, hake, cheese, baby and infant food	Food safety without artificial preservatives and additives, high nutritional quality, fresh like organoleptic quality, extending the shelf life of the product, enhancement of maturation	Suja, Pulmuone, in 2 food, Hey day, Fruselva, fruity line, Coldpress, AMC juices, Avomix, chic foods, ripe liquid, LA CASA DEL JUGO, Beskyd	Hiperbaric, (Miami, Florida, USA) Avure (Middletown, Ohio, USA)
Pulsed electric fields	Potatoes, sweet potatoes, carrots and cassava, juices and smoothies, dairy products, dried foods, liquid egg products, salad dressing	Restructure raw materials, better production planning, improved quality, bacterial spore inactivation, improved color, low temperature processing, extended market reach, retention of fresh taste, preservation of higher nutritional value, energy saving, extraction of valuable compounds, color extraction, faster extraction, accelerated drying, better sensory attributes, larger volume, better shape, faster moisture release, less shrinkage, in drying, acceleration of brining and marinating in meat products	McCain foods	Elea (Madrid, Spain) Pulsemaster (Bladel, The Netherlands)
Irradiation	Fresh and frozen meat, meat products, poultry, shellfish, fruits and vegetables, sprouts, tubers, natural gums, prepared food, wheat flour, herbs and spices, shell eggs, sea food, grains, cereals and pulses	Disinfestation of foods, disinfestation of spices, insect population control, consumer safety through pathogen reduction, inhibition of sprouting	Krushak foods, Huiskens meats, Omaha steaks, Schwan's food service	Nordion (Ottawa, Canada), Steris (Mentor, Ohio, U.S.A), Gray Star Inc. (Mt. Arlington, New Jersey, USA)

(continued)

Table 2.6 (continued)

Process	Product	Objectives	Food manufacturers	Equipment supplier
Ozonation	Meat and poultry products, seafood, sushi, breweries and wineries, fruits and vegetables, ice	Pest management, food storage, seafood processing, meat and poultry production and processing, Washing fruits and vegetables, cleaning in place (CIP) applications, surface sanitation	Tyson foods	Ozono Elettronica Internazionale (Milan, Italy), Spartan environmental technologies (Beachwood, Ohio, USA)
Supercritical CO$_2$		Decaffeination, edible oil extraction, Cork treatment,	Diam	Natex (Ternitz, Austria)
Ohmic	Pasteurized liquid eggs, dairy products	Rapid pasteurization and sterilization, improvements in the fermentation process	Campbell	C-Tech innovation (Chester, United Kingdom)
Microwave	Fried snacks, puff snacks, malt, green tea, modified starch, caramel, muesli, meat, bacon, fish, fruits and vegetables	Drying, curing, heating, cooking thawing, dry sterilization, proofing of bakery products	TOPS foods, Profood	Massalfa microwave (Jinan, China), Cellencor (Ankeny, Iowa, USA)
Infrared	Nuts, grains,	Baking, cooking, dehydrating, drying, melting, toasting, roasting, blanching, peeling,	–	Ceramicx (Cork, Ireland),
Radio frequency	Bakery products, snacks, nuts, fruits, vegetables, poultry, meat, fish	Baking and post-baking drying, rapid thawing and tempering, pasteurization, disinfestation and sanitization of nuts, cereals, grains, pulses, seeds	–	Stalam (Nove, Italy), Strayfield (Reading, United Kingdom), Litzler company (Cleveland, Ohio, USA)

References

Achir, N., Dhuique-Mayer, C., Hadjal, T., Madani, K., Pain, J. P., & Dornier, M. (2016). Pasteurization of citrus juices with ohmic heating to preserve the carotenoid profile. *Innovative Food Science & Emerging Technologies, 33*, 397–404.

ADA (2000). Position of The American Dietetic Association. *Journal of the American Dietetic Association, 100*(2), 246–253.

Adak, N., Heybeli, N., & Ertekin, C. (2017). Infrared drying of strawberry. *Food Chemistry, 219*, 109–116.

Alegbeleye, O. O., Guimarães, J. T., Cruz, A. G., & Sant'Ana, A. S. (2018). Hazards of a 'healthy'trend? An appraisal of the risks of raw milk consumption and the potential of novel treatment technologies to serve as alternatives to pasteurization. *Trends in Food Science & Technology, 82*, 148–166.

Alfaifi, B., Tang, J., Rasco, B., Wang, S., & Sablani, S. (2016). Computer simulation analyses to improve radio frequency (RF) heating uniformity in dried fruits for insect control. *Innovative Food Science & Emerging Technologies, 37*, 125–137.

Alsager, O. A., Alnajrani, M. N., Abuelizz, H. A., & Aldaghmani, I. A. (2018). Removal of antibiotics from water and waste milk by ozonation: Kinetics, byproducts, and antimicrobial activity. *Ecotoxicology and Environmental Safety, 158*, 114–122.

Amaral, G. V., Silva, E. K., Cavalcanti, R. N., Martins, C. P., Andrade, L. G. Z., Moraes, J., Alvarenga, V. O., Guimarães, J. T., Esmerino, E. A., Freitas, M. Q., & Silva, M. C. (2018). Whey-grape juice drink processed by supercritical carbon dioxide technology: Physicochemical characteristics, bioactive compounds and volatile profile. *Food Chemistry, 239*, 697–703.

Anderson, A. K., & Finkelstein, R. (1919). A study of the electropure process of treating milk. *Journal of Dairy Science, 2*(5), 374–406.

Barbosa-Canovas, G. V., Albaali, A. G., Juliano, P., & Knoerzer, K. (2011). Introduction to innovative food processing technologies: Background, advantages, issues and need for multiphysics modeling. In *Innovative food processing technologies: Advances in Multiphysics simulation* (pp. 3–23). UK: IFT Press, Wiley-Blackwell.

Behrens, J. H., Barcellos, M. N., Frewer, L. J., Nunes, T. P., & Landgraf, M. (2009). Brazilian consumer views on food irradiation. *Innovative Food Science & Emerging Technologies, 10*(3), 383–389.

Blackburn, C. (2017). *Food irradiation technologies: Concepts, applications and outcomes* (Vol. 4). London: Royal Society of Chemistry.

Calado, T., Fernández-Cruz, M. L., Verde, S. C., Venâncio, A., & Abrunhosa, L. (2018). Gamma irradiation effects on ochratoxin A: Degradation, cytotoxicity and application in food. *Food Chemistry, 240*, 463–471.

Carreño, I. (2017). International standards and regulation on food irradiation. In I. C. F. R. Ferreira, A. L. Antonio, & S. C. Verde (Eds.), *Food irradiation technologies: Concepts, applications and outcomes* (pp. 5–27). London: Royal Society of Chemistry

Castanha, N., da Matta Junior, M. D., & Augusto, P. E. D. (2017). Potato starch modification using the ozone technology. *Food Hydrocolloids, 66*, 343–356.

Cho, W. I., Yi, J. Y., & Chung, M. S. (2016). Pasteurization of fermented red pepper paste by ohmic heating. *Innovative Food Science & Emerging Technologies, 34*, 180–186.

Choi, E. J., Yang, H. S., Park, H. W., & Chun, H. H. (2018). Inactivation of Escherichia coli O157: H7 and Staphylococcus aureus in red pepper powder using a combination of radio frequency thermal and indirect dielectric barrier discharge plasma non-thermal treatments. *LWT, 93*, 477–484.

Chung, C. C., Huang, T. C., Li, C. Y., & Chen, H. H. (2013). Agriproducts sterilization and optimization by using supercritical carbon dioxide fluid (SC-CO2). In *4th International Conference on Food Engineering and Biotechnology Singapore* (pp. 1–8).

Cullen, P. J., Tiwari, B. K., & Valdramidis, V. P. (2012). Status and trends of novel thermal and non-thermal technologies for fluid foods. In *Novel thermal and non-thermal technologies for fluid foods* (pp. 1–6). Cambridge: Academic press.

De Silva, G. O., Abeysundara, A. T., Minoli, M., & Aponso, W. (2018). Impacts of pulsed electric field (PEF) technology in different approaches of food industry: A review. *Journal of Pharmacognosy and Phytochemistry, 7*(2), 737–740.

de Souza, L. P., Faroni, L. R. D. A., Heleno, F. F., Cecon, P. R., Gonçalves, T. D. C., da Silva, G. J., & Prates, L. H. F. (2018). Effects of ozone treatment on postharvest carrot quality. *LWT, 90*, 53–60.

de Toledo Guimarães, J., Silva, E. K., de Freitas, M. Q., de Almeida Meireles, M. A., & da Cruz, A. G. (2018). Non-thermal emerging technologies and their effects on the functional properties of dairy products. *Current Opinion in Food Science, 22*, 62–66.

Deeth, H. C., & Lewis, M. J. (2017). *High temperature processing of milk and milk products.* Hoboken: Wiley.

del Valle, J. M. (2015). Extraction of natural compounds using supercritical CO_2: Going from the laboratory to the industrial application. *The Journal of Supercritical Fluids, 96*, 180–199.

Derrien, M., Aghabararnejad, M., Gosselin, A., Desjardins, Y., Angers, P., & Boumghar, Y. (2018). Optimization of supercritical carbon dioxide extraction of lutein and chlorophyll from spinach by-products using response surface methodology. *LWT, 93*, 79–87.

Di Giacomo, G., Scimia, F., & Taglieri, L. (2016). Application of supercritical carbon dioxide for the preservation of fresh-like carrot juice. *International Journal of New Technology and Research, 2*(2), 71–77.

Diehl, J. F. (1999). *Safety of irradiated foods.* Boca Raton: CRC Press.

Ding, C., Khir, R., Pan, Z., Zhao, L., Tu, K., El-Mashad, H., & McHugh, T. H. (2015). Improvement in shelf life of rough and brown rice using infrared radiation heating. *Food and Bioprocess Technology, 8*(5), 1149–1159.

Doona, C. J. (Ed.). (2010). *Case studies in novel food processing technologies: Innovations in processing, packaging, and predictive modelling.* London: Elsevier.

Duranton, F., Simonin, H., Guyon, C., Jung, S., & de Lamballerie, M. (2014). High-pressure processing of meats and seafood. In *Emerging technologies for food processing* (pp. 35–63). Cambridge: Academic press.

Erdogdu, F., Altin, O., Marra, F., & Bedane, T. F. (2017). A computational study to design process conditions in industrial radio-frequency tempering/thawing process. *Journal of Food Engineering, 213*, 99–112.

Eskandari, J., Kermani, A. M., Kouravand, S., & Zarafshan, P. (2018). Design, fabrication, and evaluation a laboratory dry-peeling system for hazelnut using infrared radiation. *LWT, 90*, 570–576.

Evans, G., & Cox, D. N. (2006). Australian consumers' antecedents of attitudes towards foods produced by novel technologies. *British Food Journal, 108*(11), 916–930.

Farahnaky, A., Azizi, R., & Gavahian, M. (2012). Accelerated texture softening of some root vegetables by ohmic heating. *Journal of Food Engineering, 113*(2), 275–280.

FDA. (2009). Chapter IV. Outbreaks Associated with Fresh and Fresh-Cut Produce. Incidence, Growth, and Survival of Pathogens in Fresh and Fresh-Cut Produce. http://www.fda.gov/Food/ScienceResearch/ResearchAreas/SafePracticesforFoodProcesses/ucm091265.htm/

Fundo, J. F., Miller, F. A., Tremarin, A., Garcia, E., Brandão, T. R., & Silva, C. L. (2018). Quality assessment of Cantaloupe melon juice under ozone processing. *Innovative Food Science & Emerging Technologies, 47*, 461–466.

GAO, (2000). Food Irradiation: Available Research Indicates that Benefits Outweigh Risks. U.S. General Accounting Office, GAO/RCED-00-217. http://www.gao.gov/archive/2000/rc00217.pdf/

García-Risco, M. R., Hernández, E. J., Vicente, G., Fornari, T., Señoráns, F. J., & Reglero, G. (2011). Kinetic study of pilot-scale supercritical CO_2 extraction of rosemary (Rosmarinus officinalis) leaves. *The Journal of Supercritical Fluids, 55*(3), 971–976.

Gili, R. D., Palavecino, P. M., Penci, M. C., Martinez, M. L., & Ribotta, P. D. (2017). Wheat germ stabilization by infrared radiation. *Journal of Food Science and Technology, 54*(1), 71–81.

Gomes, C. F., Sarkis, J. R., & Marczak, L. D. F. (2018). Ohmic blanching of Tetsukabuto pumpkin: Effects on peroxidase inactivation kinetics and color changes. *Journal of Food Engineering, 233*, 74–80.

Grahl, T., & Märkl, H. (1996). Killing of microorganisms by pulsed electric fields. *Applied Microbiology and Biotechnology, 45*(1–2), 148–157.

Granella, S. J., Christ, D., Werncke, I., Bechlin, T. R., & Coelho, S. R. M. (2018). Effect of drying and ozonation process on naturally contaminated wheat seeds. *Journal of Cereal Science, 80*, 205–211.

Hu, G., Zheng, Y., Liu, Z., Xiao, Y., Deng, Y., & Zhao, Y. (2017). Effects of high hydrostatic pressure, ultraviolet light-C, and far-infrared treatments on the digestibility, antioxidant and antihypertensive activity of α-casein. *Food Chemistry, 221*, 1860–1866.

Huang, Z., Zhang, B., Marra, F., & Wang, S. (2016). Computational modelling of the impact of polystyrene containers on radio frequency heating uniformity improvement for dried soybeans. *Innovative Food Science & Emerging Technologies, 33*, 365–380.

Hunt, N. K., & Mariñas, B. J. (1999). Inactivation of Escherichia coli with ozone: Chemical and inactivation kinetics. *Water Research, 33*(11), 2633–2641.

Icier, F., Yildiz, H., Sabanci, S., Cevik, M., & Cokgezme, O. F. (2017). Ohmic heating assisted vacuum evaporation of pomegranate juice: Electrical conductivity changes. *Innovative Food Science & Emerging Technologies, 39*, 241–246.

Ilgaz, S., Sat, I. G., & Polat, A. (2018). Effects of processing parameters on the caffeine extraction yield during decaffeination of black tea using pilot-scale supercritical carbon dioxide extraction technique. *Journal of Food Science and Technology, 55*(4), 1407–1415.

Irakli, M., Kleisiaris, F., Mygdalia, A., & Katsantonis, D. (2018). Stabilization of rice bran and its effect on bioactive compounds content, antioxidant activity and storage stability during infrared radiation heating. *Journal of Cereal Science, 80*, 135–142.

Jermann, C., Koutchma, T., Margas, E., Leadley, C., & Ros-Polski, V. (2015). Mapping trends in novel and emerging food processing technologies around the world. *Innovative Food Science & Emerging Technologies, 31*, 14–27.

Jiang, Y., Wang, S., He, F., Fan, Q., Ma, Y., Yan, W., Tang, Y., Yang, R., & Zhao, W. (2018). Inactivation of lipoxygenase in soybean by radio frequency treatment. *International Journal of Food Science & Technology, 53*(12), 2738–2747.

Kalla, A. M. (2017). Microwave energy and its application in food industry: A review. *Asian Journal of Dairy & Food Research, 36*(1), 37–44.

Kanjanapongkul, K. (2017). Rice cooking using ohmic heating: Determination of electrical conductivity, water diffusion and cooking energy. *Journal of Food Engineering, 192*, 1–10.

Kempkes, M., Simpson, R., & Roth, I. (2016). Removing barriers to commercialization of PEF systems and processes. In *Proceedings of 3rd School on Pulsed Electric Field Processing of Food* (pp. 1–6). Dublin: Institute of Food and Health, University College Dublin.

Kettler, K., Adhikari, K., & Singh, R. K. (2017). Blanchability and sensory quality of large runner peanuts blanched in a radiant wall oven using infrared radiation. *Journal of the Science of Food and Agriculture, 97*(13), 4621–4628.

Khadre, M. A., Yousef, A. E., & Kim, J. G. (2001). Microbiological aspects of ozone applications in food: A review. *Journal of Food Science, 66*(9), 1242–1252.

Kim, J. G., Yousef, A. E., & Dave, S. (1999). Application of ozone for enhancing the microbiological safety and quality of foods: A review. *Journal of Food Protection, 62*(9), 1071–1087.

Kim, S. Y., Sagong, H. G., Choi, S. H., Ryu, S., & Kang, D. H. (2012). Radio-frequency heating to inactivate Salmonella Typhimurium and Escherichia coli O157: H7 on black and red pepper spice. *International Journal of Food Microbiology, 153*(1–2), 171–175.

Kirmaci, B., & Singh, R. K. (2012). Quality of chicken breast meat cooked in a pilot-scale radio frequency oven. *Innovative Food Science & Emerging Technologies, 14*, 77–84.

Knez, Ž., Markočič, E., Leitgeb, M., Primožič, M., Hrnčič, M. K., & Škerget, M. (2014). Industrial applications of supercritical fluids: A review. *Energy, 77*, 235–243.

Knoerzer, K., Buckow, R., Trujillo, F. J., & Juliano, P. (2015). Multiphysics simulation of innovative food processing technologies. *Food Engineering Reviews, 7*(2), 64–81.

Lascorz, D., Torella, E., Lyng, J. G., & Arroyo, C. (2016). The potential of ohmic heating as an alternative to steam for heat processing shrimps. *Innovative Food Science & Emerging Technologies, 37*, 329–335.

Leadley, C. (2008). Novel commercial preservation methods. In G. S. Tucker (Ed.), *Food biodeterioration and preservation* (pp. 211–242). Hoboken: Blackwell Publishing.

Liao, M., Zhao, Y., Gong, C., Zhang, H., & Jiao, S. (2018). Effects of hot air-assisted radio frequency roasting on quality and antioxidant activity of cashew nut kernels. *LWT, 93*, 274–280.

Ling, B., Lyng, J. G., & Wang, S. (2018). Effects of hot air-assisted radio frequency heating on enzyme inactivation, lipid stability and product quality of rice bran. *LWT, 91*, 453–459.

Luo, X., Wang, R., Wang, L., Li, Y., Bian, Y., & Chen, Z. (2014). Effect of ozone treatment on aflatoxin B1 and safety evaluation of ozonized corn. *Food Control, 37*, 171–176.

Mahapatra, A. K., Muthukumarappan, K., & Julson, J. L. (2005). Applications of ozone, bacteriocins and irradiation in food processing: A review. *Critical Reviews in Food Science and Nutrition, 45*(6), 447–461.

Mans, J., & Swientek, B. (1993). Electrifying progress in aseptic technology. *Prepared foods, 162*(9), 151–156.

Meda, V., Orsat, V., & Raghavan, V. (2017). Microwave heating and the dielectric properties of foods. In *The microwave processing of foods* (pp. 23–43). Cambridge: Woodhead Publishing.

Michelino, F., Zambon, A., Vizzotto, M. T., Cozzi, S., & Spilimbergo, S. (2018). High power ultrasound combined with supercritical carbon dioxide for the drying and microbial inactivation of coriander. *Journal of CO2 Utilization, 24*, 516–521.

Mittendorfer, J. (2016). Food irradiation facilities: Requirements and technical aspects. *Radiation Physics and Chemistry, 129*, 61–63.

Muntean, M. V., Marian, O., Barbieru, V., Cătunescu, G. M., Ranta, O., Drocas, I., & Terhes, S. (2016). High pressure processing in food industry–characteristics and applications. *Agriculture and Agricultural Science Procedia, 10*, 377–383.

Muthukumarappan, K. (2011). Ozone processing. In D. W. Sun (Ed.) *Handbook of food safety engineering* (pp. 681–692), 681–692. West Sussex: Wiley-Blackwell.

Muthukumarappan, K., Halaweish, F., & Naidu, A. S. (2000). Ozone. In A. S. Naidu (Ed.), *Natural food antimicrobial systems* (pp. 796–813). Boca Raton: CRC Press.

Muthukumarappan, K., Julson, J. L., Mahapatra, A. K., & Nanda, S. K. (2002). Ozone applications in food processing. In *Souvenir 2002—-Proc. College of Agric. Eng. Technol. alumnai meet* (pp. 32–35).

Nagaraj, G., Purohit, A., Harrison, M., Singh, R., Hung, Y. C., & Mohan, A. (2016). Radiofrequency pasteurization of inoculated ground beef homogenate. *Food Control, 59*, 59–67.

Niemira, B. A., & Gao, M. (2012). Irradiation of fluid foods. In P.J. Cullen, B.K. Tiwari, V. Valdramidis (Eds.), *Novel thermal and non-thermal Technologies for Fluid Foods* (pp. 167–183). Cambridge: Academic press.

Omar, A. M., Norsalwani, T. T., Asmah, M. S., Badrulhisham, Z. Y., Easa, A. M., Omar, F. M., Hossain, M. S., Zuknik, M. H., & Norulaini, N. N. (2018). Implementation of the supercritical carbon dioxide technology in oil palm fresh fruits bunch sterilization: A review. *Journal of CO2 Utilization, 25*, 205–215.

Oualid, H. A., Amadine, O., Essamlali, Y., Dânoun, K., & Zahouily, M. (2018). Supercritical CO_2 drying of alginate/zinc hydrogels: A green and facile route to prepare ZnO foam structures and ZnO nanoparticles. *RSC Advances, 8*(37), 20737–20747.

Ozturk, S., Kong, F., Trabelsi, S., & Singh, R. K. (2016). Dielectric properties of dried vegetable powders and their temperature profile during radio frequency heating. *Journal of Food Engineering, 169*, 91–100.

Ozturk, S., Kong, F., Singh, R. K., Kuzy, J. D., Li, C., & Trabelsi, S. (2018). Dielectric properties, heating rate, and heating uniformity of various seasoning spices and their mixtures with radio frequency heating. *Journal of Food Engineering, 228*, 128–141.

Palazoğlu, T. K., & Miran, W. (2018). Experimental investigation of the effect of conveyor movement and sample's vertical position on radio frequency tempering of frozen beef. *Journal of Food Engineering, 219*, 71–80.

Patil, S., & Bourke, P. (2012). Ozone processing of fluid foods. In *Novel thermal and non-thermal technologies for fluid foods* (pp. 225–261). Cambridge: Academic press.

Pawar, S. B., & Pratape, V. M. (2017). Fundamentals of infrared heating and its application in drying of food materials: A review. *Journal of Food Process Engineering, 40*(1), e12308.

Pereira, R. N., & Vicente, A. A. (2010). Environmental impact of novel thermal and non-thermal technologies in food processing. *Food Research International, 43*(7), 1936–1943.

Pereira, R. N., Rodrigues, R. M., Genisheva, Z., Oliveira, H., de Freitas, V., Teixeira, J. A., & Vicente, A. A. (2016). Effects of ohmic heating on extraction of food-grade phytochemicals from colored potato. *LWT, 74*, 493–503.

Perrut, M. (2012). Sterilization and virus inactivation by supercritical fluids (a review). *The Journal of Supercritical Fluids, 66*, 359–371.

Rahman, M. S., Seo, J. K., Choi, S. G., Gul, K., & Yang, H. S. (2018). Physicochemical characteristics and microbial safety of defatted bovine heart and its lipid extracted with supercritical-CO_2 and solvent extraction. *LWT, 97*, 355–361.

Ramaswamy, R., Jin, T., Balasubramaniam, V. M., & Zhang, H. (2005). Pulsed electric field processing: fact sheet for food processors. *Ohio State University Extension Factsheet*, 22.

Rawson, A., Tiwari, B. K., Brunton, N., Brennan, C., Cullen, P. J., & O'donnell, C. P. (2012). Application of supercritical carbon dioxide to fruit and vegetables: Extraction, processing, and preservation. *Food Reviews International, 28*(3), 253–276.

Riadh, M. H., Ahmad, S. A. B., Marhaban, M. H., & Soh, A. C. (2015). Infrared heating in food drying: An overview. *Drying Technology, 33*(3), 322–335.

Rice, R. G., Overbeck, P., & Larson, K. A. (2000). Costs of ozone in small drinking water systems. In *Proc. Small drinking water and wastewater systems*. Ann Arbor: NSF International.

Rincon, A. M., Singh, R. K., & Stelzleni, A. M. (2015). Effects of endpoint temperature and thickness on quality of whole muscle non-intact steaks cooked in a radio frequency oven. *LWT-Food Science and Technology, 64*(2), 1323–1328.

Ruan, R., Ye, X., Chen, P., Doona, C. J., Taub, I., & Center, N. S. (2001). Ohmic heating. In P. Richardson (Ed.), *Thermal technologies in food processing* (pp. 241–265). Boca Raton: CRC Press.

Sabanci, S., & Icier, F. (2017). Applicability of ohmic heating assisted vacuum evaporation for concentration of sour cherry juice. *Journal of Food Engineering, 212*, 262–270.

Saberian, H., Hamidi-Esfahani, Z., Gavlighi, H. A., & Barzegar, M. (2017). Optimization of pectin extraction from orange juice waste assisted by ohmic heating. *Chemical Engineering and Processing: Process Intensification, 117*, 154–161.

Sarkis, J. R., Jaeschke, D. P., Tessaro, I. C., & Marczak, L. D. (2013). Effects of ohmic and conventional heating on anthocyanin degradation during the processing of blueberry pulp. *LWT-Food Science and Technology, 51*(1), 79–85.

Schlisselberg, D. B., Kler, E., Kalily, E., Kisluk, G., Karniel, O., & Yaron, S. (2013). Inactivation of foodborne pathogens in ground beef by cooking with highly controlled radio frequency energy. *International Journal of Food Microbiology, 160*(3), 219–226.

Smith, J. S., & Pillai, S. (2004). Irradiation and food safety. *Food Technology, 58*(11), 48–55.

Soares, V. B., & Coelho, G. L. (2012). Safety study of an experimental apparatus for extraction with supercritical CO2. *Brazilian Journal of Chemical Engineering, 29*(3), 677–682.

Subramaniam, B. (2017). Sustainable processes with supercritical fluids. In: Encyclopedia of Sustainable Technologies, Pages 653–662. Elsevier, UK.

Tello, J., Viguera, M., & Calvo, L. (2011). Extraction of caffeine from Robusta coffee (Coffea canephora var. Robusta) husks using supercritical carbon dioxide. *The Journal of Supercritical Fluids, 59*, 53–60.

Termrittikul, P., Jittanit, W., & Sirisansaneeyakul, S. (2018). The application of ohmic heating for inulin extraction from the wet-milled and dry-milled powders of Jerusalem artichoke (Helianthus tuberosus L.) tuber. *Innovative Food Science & Emerging Technologies, 48*, 99–110.

Thayer, D. W., & Boyd, G. (1995). Radiation sensitivity of Listeria monocytogenes on beef as affected by temperature. *Journal of Food Science, 60*(2), 237–240.

Tiwari, G., Wang, S., Tang, J., & Birla, S. L. (2011). Computer simulation model development and validation for radio frequency (RF) heating of dry food materials. *Journal of Food Engineering, 105*(1), 48–55.

Tuncel, N. B., Uygur, A., & Yüceer, Y. K. (2017). The effects of infrared roasting on HCN content, chemical composition and storage stability of flaxseed and flaxseed oil. *Journal of the American Oil Chemists' Society, 94*(6), 877–884.

Turp, G. Y., Icier, F., & Kor, G. (2016). Influence of infrared final cooking on color, texture and cooking characteristics of ohmically pre-cooked meatball. *Meat Science, 114*, 46–53.

Valadez-Carmona, L., Ortiz-Moreno, A., Ceballos-Reyes, G., Mendiola, J. A., & Ibáñez, E. (2018). Valorization of cacao pod husk through supercritical fluid extraction of phenolic compounds. *The Journal of Supercritical Fluids, 131*, 99–105.

Van Bockstal, P. J., De Meyer, L., Corver, J., Vervaet, C., & De Beer, T. (2017). Noncontact infrared-mediated heat transfer during continuous freeze-drying of unit doses. *Journal of Pharmaceutical Sciences, 106*(1), 71–82.

Venkitasamy, C., Zhu, C., Brandl, M. T., Niederholzer, F. J., Zhang, R., McHugh, T. H., & Pan, Z. (2018). Feasibility of using sequential infrared and hot air for almond drying and inactivation of enterococcus faecium NRRL B-2354. *LWT, 95*, 123–128.

Wang, S., Tiwari, G., Jiao, S., Johnson, J. A., & Tang, J. (2010). Developing postharvest disinfestation treatments for legumes using radio frequency energy. *Biosystems Engineering, 105*(3), 341–349.

Wang, Y., Zhang, L., Johnson, J., Gao, M., Tang, J., Powers, J. R., & Wang, S. (2014). Developing hot air-assisted radio frequency drying for in-shell macadamia nuts. *Food and Bioprocess Technology, 7*(1), 278–288.

Wang, H. C., Zhang, M., & Adhikari, B. (2015). Drying of shiitake mushroom by combining freeze-drying and mid-infrared radiation. *Food and Bioproducts Processing, 94*, 507–517.

Wongsa-Ngasri, P., & Sastry, S. K. (2015). Effect of ohmic heating on tomato peeling. *LWT-Food Science and Technology, 61*(2), 269–274.

World Health Organization (WHO), 1999. High-Dose Irradiation: Wholesomeness of Food Irradiated with Doses above 10 KGy, Joint FAO/IAEA/WHO Study Group on HighDose Irradiation, Geneva, 1520 September 1997, WHO Technical Report Series 890

Xie, L., Cahoon, E., Zhang, Y., & Ciftci, O. N. (2019). Extraction of astaxanthin from engineered Camelina sativa seed using ethanol-modified supercritical carbon dioxide. *The Journal of Supercritical Fluids, 143*, 171–178.

Xu, L. (1999). Use of ozone to improve the safety of fresh fruits and vegetables. *Food Technology, 53*, 58–63.

Yalcin, S., & Basman, A. (2015). Effects of infrared treatment on urease, trypsin inhibitor and lipoxygenase activities of soybean samples. *Food Chemistry, 169*, 203–210.

Yılmaz, N. (2016). Middle infrared stabilization of individual rice bran milling fractions. *Food Chemistry, 190*, 179–185.

Zabot, G. L., Moraes, M. N., Petenate, A. J., & Meireles, M. A. A. (2014). Influence of the bed geometry on the kinetics of the extraction of clove bud oil with supercritical CO2. *The Journal of Supercritical Fluids, 93*, 56–66.

Zhang, X., Heinonen, S., & Levänen, E. (2014). Applications of supercritical carbon dioxide in materials processing and synthesis. *RSC Advances, 4*(105), 61137–61152.

Zhang, Z. S., Xie, Q. F., & Che, L. M. (2018). Effects of gamma irradiation on aflatoxin B1 levels in soybean and on the properties of soybean and soybean oil. *Applied Radiation and Isotopes, 139,* 224–230.

Zheng, A., Zhang, B., Zhou, L., & Wang, S. (2016). Application of radio frequency pasteurization to corn (Zea mays L.): Heating uniformity improvement and quality stability evaluation. *Journal of Stored Products Research, 68,* 63–72.

Zheng, A., Zhang, L., & Wang, S. (2017). Verification of radio frequency pasteurization treatment for controlling Aspergillus parasiticus on corn grains. *International Journal of Food Microbiology, 249,* 27–34.

Zhou, L., Ling, B., Zheng, A., Zhang, B., & Wang, S. (2015). Developing radio frequency technology for postharvest insect control in milled rice. *Journal of Stored Products Research, 62,* 22–31.

Zhu, F. (2018). Effect of ozone treatment on the quality of grain products. *Food Chemistry, 264,* 358.

Zhu, H., Li, D., Ma, J., Du, Z., Li, P., Li, S., & Wang, S. (2018a). Radio frequency heating uniformity evaluation for mid-high moisture food treated with cylindrical electromagnetic wave conductors. *Innovative Food Science & Emerging Technologies, 47,* 56–70.

Zhu, X. H., Yang, Y. X., & Duan, Z. H. (2018b). Research progress on the effect of microwave sterilization on agricultural products quality. In *IOP Conference Series: Earth and Environmental Science* (Vol. 113, No. 1, p. 012096). Bristol: IOP Publishing.

Chapter 3
Post-Harvest Treatments and Related Food Quality

Bernhard Trierweiler and Christoph H. Weinert

3.1 Introduction

Fruit and vegetables are still living plant organs after harvest with an active metabolism and ongoing respiration, ripening, and senescence processes which have to be controlled to maintain the quality of the products (Brasil and Siddiqui 2018). Post-harvest losses of, e.g., apples due to ripening and respiration may add up to 4–10% but can be reduced to 2–6% by appropriate post-harvest treatments (Roser et al. 2013). Quality of fruit and vegetables cannot be improved after harvest, it can only be preserved (Fallik and Ilic 2018). Therefore, the preservation of valuable compounds like vitamins and secondary metabolites is a main goal in post-harvest treatments. Furthermore, sensory parameters like appearance, texture, and flavour are important for acceptability by consumers and must be taken into account for post-harvest treatments (Aked 2002).

As fruit and vegetables are preferentially consumed at full ripeness, their sensory quality is mainly tested at this stage. However, certain fruits are typically harvested in an unripe stage for consumption (e.g., banana, kiwifruit, mango, and avocado) in order to enable shipment over long distances, often between continents. Such fruits have to be after-ripened in the country of destination (for example by an increase in temperature and an ethylene treatment) before marketing. While this is unavoidable from the economic point of view, the desired quality may be difficult to achieve by after-ripening of prematurely harvested fruit (Karapanos et al. 2015). On the one hand, kiwi fruit stored under controlled atmosphere and ripened after storage possessed a similar sensory quality as vine ripened fruit (Latocha et al. 2014). On the other hand, mango harvested in an immature status were found to be more sensitive to chilling injuries and may fail to ripen properly (Sivakumar et al. 2011). Further,

B. Trierweiler (✉) · C. H. Weinert
Department of Safety and Quality of Fruit and Vegetables, Federal Research Institute
of Nutrition and Food, Max Rubner-Institut (MRI), Karlsruhe, Germany
e-mail: Bernhard.Trierweiler@mri.bund.de; christoph.weinert@mri.bund.de

© Springer Nature Switzerland AG 2019
C. Piatti et al. (eds.), *Food Tech Transitions*,
https://doi.org/10.1007/978-3-030-21059-5_3

the aroma profile changes significantly during ripening (Obenland et al. 2012) and this process may also be influenced by the time of harvest. Again, it has to be emphasized that quality of fruit and vegetables is determined during cultivation and can only be preserved, but not improved after harvest.

The most important factor to maintain fruit and vegetable quality is the appropriate temperature during storage and transport. For example, preservation of anthocyanins, an important group of secondary metabolites in fruit and vegetables, is very much influenced by temperature (Raffo et al. 2008; Odriozola-Serrano et al. 2009). Besides temperature, the composition of storage atmosphere is another important factor influencing respiration and in conjunction ripening and senescence of fruit and vegetables (Harman and McDonald 1989; Simões et al. 2011; Bekele et al. 2016).

Appropriate post-harvest storage and transport conditions are not always available especially in developing countries. Therefore, further post-harvest treatments in developing countries like Africa and Asia, e.g. hot water treatment, modified atmosphere packaging, and fermentation are necessary to maintain food security and world nutrition (Habwe et al. 2008; Wafula et al. 2016; Ndlela et al. 2017). The different post-harvest treatments described in the following sections are useable globally and not only in developing countries.

3.2 Post-Harvest Treatments and Food Quality

3.2.1 Methods for Quality Determination

The first quality determination has to be done before harvest. To determine the appropriate harvest time at an optimal product specific ripeness, the following parameters are evaluated: firmness, total soluble solid (°Brix), and starch content. These parameters together resulting in the so called Streif-Index for ripeness, especially for apples (Winter and Link 2002). After harvest, a larger set of basic chemical or enzymatic assays is traditionally performed in order to describe fruit quality status more comprehensively. In addition to the aforementioned parameters, in part depending on the kind of fruit or vegetable of interest, also vitamin C content, total polyphenols, reducing sugars, total acidity etc. are determined (Table 3.1). Furthermore, sensory testing by a trained panel can be an additional method for quality evaluation especially to determine consumer acceptance after post-harvest treatments. All the aforementioned analytical methods target specific compounds, compound classes or chemical sum parameters and provide thus valuable information about basic quality characteristics. However, recent technological advances have enabled the development of extremely powerful analytical instruments which are able to detect and quantify hundreds to thousands of known and unknown compounds in one analysis. This facilitated the wide-spread application of the untargeted metabolomics approach also in the post-harvest field which enables an in-depth evaluation of fruit composition and thus fruit quality (Nicolaï et al. 2010;

Table 3.1 Methods and assays used to determine basic fruit quality characteristics

Measured parameter	Analytical principle/methodology	Detected property or chemical species. Possible limitations	Literature
Fruit colour	(a) Visual comparison with colour charts (rather subjective) (b) Instruments measuring the composition of light after reflection or transmission	Sum of all natural dyes, e.g., anthocyanins, carotenoids, betalains, chlorophyll, riboflavin and others	Mitcham et al. (1996/2003)
Gloss	Glossmeters measuring specular gloss or other types of gloss	The ability of fruit surface to reflect light without scattering	Mizrach et al. (2009)
Defects and disorders	Visual evaluation. Defects are counted and classified according to their severity.	Cuts, bruises, scald, bitter pit, internal browning, chilling injuries, etc.	Mitcham et al. (1996/2003)
Firmness	Handheld or benchtop penetrometers measuring, for example, the pressure needed to push a plunger of defined size through the peel into the pulp up to a certain depth	Firmness or degree of softness/crispness	Mitcham et al. (1996/2003) and OECD (2009)
Juice content	Gravimetric determination of the juice yield (in %) using an extractor or a juice press		OECD (2009)
Water content and dry matter (total solids)	Drying in a ventilated or vacuum oven, in case of fruit and vegetables usually at 60–70 °C	Water content and the residual dry matter	Mauer and Bradley (2017)
Total soluble solids	Determination of the refractive index (°Brix) using a refractometer	Mainly sugars derived from starch during ripening. Other compounds like organic acids, amino acids and polyphenols may also contribute significantly	Mitcham et al. (1996/2003) and OECD (2009)
pH	Potentiometric analysis using a pH meter (in juices or extracts)	Concentration (or activity) of H_3O^+ ions	Tyl and Sadler (2017)
Titratable acidity	Titration (of liquid foods or extracts) with a strong base (e.g., sodium hydroxide) until a target pH of 8.1 is reached or a colour change of a pH-sensitive dye occurs. Result is expressed as % or g/L of a reference acid	Total acid concentration including all organic (free and bound) and inorganic acids	Tyl and Sadler (2017)

(continued)

Table 3.1 (continued)

Measured parameter	Analytical principle/ methodology	Detected property or chemical species. Possible limitations	Literature
Total phenolics	(a) Photometric determination after reaction of phenolics with the Folin-Ciocalteu reagent.	Simple phenols and polyphenols. Gallic acid is used as reference for quantification. Limited selectivity, high amounts of, e.g., thiols, reducing sugars, amino acids or ascorbic acid may lead to biased results	Waterhouse (2002), Sánchez-Rangel et al. (2013), and Bunzel and Schendel (2017)
	(b) Direct photometric analysis		
Antioxidant capacity	Determination of the ability of antioxidants to react with radicals derived from a radical initiator in a given test system. Typically, the formation or the degradation of a photometrically or fluorimetrically active compound is measured. Quantification is performed in relation to a reference compound like Trolox	All compounds which are able to scavenge radicals derived from the radical initiator used, e.g., AAPH (ORAC assay), ABTS (TEAC assay) or DPPH (DPPH assay). Each test system has its specific limitations and comparison of results may be difficult	Karadag et al. (2009) and Bunzel and Schendel (2017)
Mono- und disaccharides	(a) Various enzymatic assays	Sucrose, D-glucose and D-fructose as well as other sugars	BeMiller (2017)
	(b) Determination using HLPC with RI or electrochemical detection		
Vitamin C	(a) Enzymatic determination	L-ascorbic acid (and dehydroascorbic acid, if reduced to ascorbic acid)	Matissek et al. (2018)
	(b) HLPC with UV detection		
Starch	(a) In-situ determination of starch content in fruit at harvest using iodine/potassium iodide solution	(a) Starch, colorized by intercalation of iodine in starch chain	OECD (2009)
	(b) Quantitative determination of starch content by an enzymatic assay, e.g. r-biopharm	(b) Starch after degradation to D-glucose	

Benkeblia 2014). Consequently, an increasing number of metabolomics studies has been performed especially in the last decade, for example to investigate pre- and post-harvest ripening or fruit development processes (Jom et al. 2011; Nardozza et al. 2013; Mack et al. 2017), to describe changes of the metabolite profile during storage (Hatoum et al. 2014; Brizzolara et al. 2017) and to elucidate the effects of technological treatments (Rudell et al. 2008; Picó et al. 2010; Leisso et al. 2013; Lopez-Sanchez et al. 2015) or the mechanisms of fruit resistance against pathogens (Wojciechowska et al. 2014). It can be expected that a broader application of metabolomics in the post-harvest field will further improve our understanding of the molecular basis of a high fruit quality.

3.2.2 Optimized Storage Conditions

3.2.2.1 Cold Storage

Fruit and vegetables have still an active metabolism and a respiration activity after harvest. These activities are strongly influenced by temperature during transport and storage. A reduction of temperature of 10 °C is slowing the metabolic activity (activity of ripening and degradation enzymes) of fruit and vegetables by a factor of two to three (Winter and Link 2002). Therefore, on the one hand, temperature during transport and storage should be at a product specific minimum to preserve harvest quality of fruit and vegetables. On the other hand, diverse fruit and vegetables have different susceptibility for chilling injuries, which means that a fruit- and in part cultivar-specific optimal temperature needs to be figured out (Table 3.2).

Table 3.2 Storability of different fruits and vegetables under cold storage conditions at normal atmosphere

Type of fruit or vegetable	Species/cultivar		Cold storage conditions		
			Temperature (°C)	Relative humidity (%)	Storage time
Pome	Apple	Berlepsch	1	93–95	4 months
		Boskoop	3–4	90–92	5–6 months
		Braeburn	1	90–95	5–6 months
		Elstar	2–3	93–95	3–4 months
		Fuji	1	93–95	5–6 months
		Gala	1	93–95	3–4 months
		Golden Delicious	1	93–95	4 months
		Ingrid Marie	1–2	90–92	3–4 months
		Jonagold	1–2	90–93	3–4 months
		Pinova	1	90–93	3–4 months
		Rubinette	−0.5 to 2	90–93	3–5 months
		Topaz	1	90–93	3–5 months
	Pear		−1 to 0	90–93	4 months
Stone fruit	Apricot		−1 to 0	95–97	3 weeks
	Cherry (sour)		−0.5 to 2	90–95	2 weeks
	Cherry (sweet)		−1	90–95	1–3 weeks
	Peach/nectarine		−1 to 0	90–95	2–5 weeks
	Plum		1	90–95	2–5 weeks
	Yellow plum		0.5–1	90	3 weeks
Berries	Currants		0–1	90	4 weeks
	Gooseberries		1	90–95	3 weeks
	Raspberries		−1 to 0	90–95	3 days
	Strawberries		0–2	90–95	5 days
	Table grapes		−0.5 to 1	95	6 weeks

(continued)

Table 3.2 (continued)

Type of fruit or vegetable	Species/cultivar	Cold storage conditions		
		Temperature (°C)	Relative humidity (%)	Storage time
Tropical fruit	Banana (green)	13	90	10 days
	Banana (ripe)	13	90	3–5 days
	Grapefruit	10–12	85–90	3 months
	Kiwi	−0.5 to 0.5	90–95	6 months
	Kumquat	10	90	4 weeks
	Lemon	12	85–90	4 months
	Lime	8–10	90	6–8 weeks
	Lychee	0–2	90–95	4–6 weeks
	Mandarin	5–6	90	6 weeks
	Mango	10–14	90	6 weeks
	Orange	8–10	85–90	4 months
	Pineapple (green)	11–12	85–90	4–5 weeks
	Pineapple (½ ripe)	7–8	85–90	3–4 weeks
Vegetables	Artichoke	−1 to 1	90–95	4 weeks
	Beans	7–8	90–95	10 days
	Broccoli	0	96–98	2 weeks
	Cabbage	0–0,5	95	6–7 months
	Carrots	0.5	98	5–6 months
	Cauliflower	0	92–95	2–3 weeks
	Chicory	0–1	90–95	4 weeks
	Chinese cabbage	0.5–1	95–98	2–3 months
	Courgette	7–10	90–95	3 weeks
	Cucumber	7–10	90–95	10–14 days
	Egg plant	10	90–95	1–2 weeks
	Horseradish	−5	95	12 months
	Pepper (red/green)	8–9	90–95	3 weeks
	Pumpkin	10	60–70	2–6 months
	Tomato (½ ripe)	12–15	85–90	3 weeks
	Tomato (¾ ripe)	8–10	80–85	1–2 weeks

Sources: Max Rubner- Institut, Department of Safety and Quality of Fruit and Vegetables; Nicolaisen-Scupin/Hansen, Leitfaden für Lagerung und Transport von Gemüse und eßbaren Früchten, 1985; Osterloh et al., Handbuch der Lebensmitteltechnologie, Lagerung von Obst und Südfrüchten, Ulmer Verlag 1996

A few commodities tolerate mild freezing, for example parsnip, onions, and garlic (Aked 2002; James et al. 2009). Apples and pears for example can be stored at temperatures of 1–2 °C without physiological damages depending on the cultivar. For more sensitive apple cultivars a storage temperature of 3 °C is recommended. Even slight minus degrees are tolerated by certain apple species like 'Rubinette' (Schirmer 2001). In contrast, tropical and subtropical fruits are more susceptible to chilling injuries. These fruits must be stored at higher temperatures, mainly above 10 °C to avoid physiological damage of the fruit flesh and the skin. The storage temperature is depending on the fruit species. For example, banana and mango should be stored at 13 °C, where they keep their quality and do not ripe too fast. Citrus fruit like oranges, limes, and lemons can be transported and stored at 10–12 °C. The optimal transport and storage temperature is also depending on the maturity of the fruit. Immature, mature, unripe, and ripe fruit show a different sensitivity to temperature (Aked 2002). Ripe tropical and subtropical fruits can be stored at lower temperatures than unripe fruits. Climacteric tropical fruit like banana, mango, and avocado can be ripened after harvest by exposure to ethylene and temperatures around 20 °C. Because of this property, these products are harvested at a less mature status for better transport and ripened at the target destination.

3.2.2.2   Controlled Atmosphere Storage

Generally, storage under controlled atmosphere helps to preserve quality of fruit and vegetables over a long time, but optimal conditions for different cultivars must be determined regarding the specific requirements of the fruit and vegetables. Table 3.3 shows optimized controlled atmosphere conditions for selected fruit and vegetables. In addition to low storage temperatures, storage under controlled atmosphere (at lower oxygen concentrations, e.g., 1–3% for apples) reduces respiration, ethylene production, and physiological activity (Bekele et al. 2016) and helps to preserve harvest quality of fruit and vegetables. In general, chilling-sensitive products can be stored at slightly higher temperatures under controlled atmosphere without losing quality. However, not all fruit and vegetables tolerate elevated carbon dioxide (CO_2) concentrations in combination with very low oxygen (O_2) concentrations. For example, the apple cultivar 'Braeburn' is prone to internal browning (Felicetti et al. 2011) at higher CO_2 concentrations and very low O_2 concentrations. Therefore, for 'Braeburn' apples a storage atmosphere of <1% CO_2 and >1.5% O_2 is recommended. CO_2 resistant cultivars can be stored at 3% CO_2 and 1% O_2. Insensitive apple cultivars can so be stored under these conditions for 6–8 months at 1 °C. In contrast, redcurrants are very resistant to high CO_2 concentrations and can be stored at 25% CO_2 with 2% O_2 for a few months to preserve their quality, e.g. outer appearance, texture, and taste. In contrast, kiwifruit can be stored at CO_2 levels between 3% and 8% without any negative influence whereas carbon dioxide levels over 14% lead to less titratable acid content and abnormal texture (Harman and McDonald 1989).

3.2.2.3   Modified Atmosphere Packaging

An increasing consumer demand for convenience products – especially minimally processed and fresh-cut fruit and vegetables free of preservatives and artificial colours (Del-Valle et al. 2009) – makes it necessary to develop appropriate technologies to maintain the fresh quality of the products even outside of CA storage

Table 3.3 Storability of fruit and vegetables under controlled atmosphere (CA)

Type of fruit or vegetable	Species/cultivar		CA storage conditions				
			Temperature (°C)	Relative humidity (%)	CO$_2$ (%)	O$_2$ (%)	Storage time
Pome	Apple	Berlepsch	1	95	3	1–2	6 months
		Boskoop	4	93–95	2	2	5–7 months
		Braeburn	1	90–95	1	1–2	6–8 months
		Elstar	2–3	95	1	3	5 months
		Fuji	1	95	3	1	7 months
		Gala	1–2	93–95	3 5	2	5 months
		Golden Delicious	1–2	95	3–5	1–2	6–8 months
		Ingrid Marie	1–2	90–93	2–3	3	5 months
		Jonagold	1–2	90–95	3–5	1–2	6–7 months
		Pinova	1	90–95	3	1	6–8 months
		Rubinette	1–2	90–95	1	2–3	4–6 months
		Topaz	1	90–95	3	1	6–8 months
	Pear		−1 to 0	95	1–3	2–3	6 months
Stone fruit	Cherry (sour)		0–2	95	15–20	2–5	4 months
	Cherry (sweet)		0–2	95	15–20	1–5	7 months
	Peach/nectarine		0	95	5–10	2–5	4–6 months
	Plum		1	90–95	25	6	5–6 months
	Yellow plum		0.5–1	90	15	Air	6 months
Berries	Currants		0–2	90–95	20	1–2	6–8 months
	Gooseberries		1	90–95	15	Air	5 weeks
	Raspberries		0–2	90–95	20–25	Air	10 days
	Strawberries		0–2	90–95	15	6	10 days
	Table grapes		−0.5 to 1	90–95	15	6	2–3 months
Tropical fruit	Banana (green)		13	90	5–8	4–5	2 months
	Banana (ripe)		13	90	6	2	2 weeks
	Grapefruit		10–15	85–90	5–10	3–10	3–4 months
	Kiwi		−0.5 to 0.5	90–95	3–5	2–3	6–9 months
	Lychee		5–7	90–95	5	3–5	6 weeks
	Pineapple (½ ripe)		8–13	85–90	5–10	2–5	3–4 weeks
Vegetables	Artichoke		1	95	5–6	2	6 weeks
	Beans		7–8	95	3–5	2	2 weeks
	Broccoli		1	96–98	2–3	2–3	4 weeks
	Cabbage		0–1	95	4–5	2–3	8 months
	Cauliflower		1	95	5	3	6 weeks
	Chicory		1–2	95	4–5	3–4	8 weeks
	Chinese cabbage		0–1	98	0,5–2	2–3	4 months
	Cucumber		7–10	95	5	2	3 weeks
	Pepper (red/green)		1	95	2–3	2–3	6 weeks
	Tomato (¾ ripe)		14–15	85	3	4	3–4 weeks

Sources: Max Rubner- Institut, Department of Safety and Quality of Fruit and Vegetables; Nicolaisen-Scupin/Hansen, Leitfaden für Lagerung und Transport von Gemüse und eßbaren Früchten, 1985; Osterloh et al. Handbuch der Lebensmitteltechnologie, Lagerung von Obst und Südfrüchten, Ulmer Verlag 1996

facilities. Because of an intensive metabolic activity and spoilage caused by micro-organisms, fresh cut fruits are very perishable.

Good agricultural production practices along with an appropriate use of temperature and modified atmosphere packaging are indispensable to preserve fruit quality by reducing spoilage and thus enabling a longer shelf life (Day 2002). Modified atmosphere in plastic bags is created by using foils with a different permeability of oxygen and carbon dioxide, e.g., to increase carbon dioxide concentration and reduce oxygen concentration. This change in atmosphere aiming to preserve product quality is depending on the respiration of the packed product with higher or lower carbon dioxide production. As an example, modified atmosphere packaging may help to preserve fresh-cut fruit and vegetables, which are especially prone to degenerative changes and a shorter shelf life (Putnik et al. 2017). Therefore, temperatures from 2 to 4 °C are recommended for storage of fresh-cut fruits to reduce spoilage, quality loss, and to improve food safety. However, storage of fresh-cut fruit and vegetables at such low temperatures is not always possible, especially during transport. To preserve food quality even at slightly higher temperatures, the use of modified atmosphere packaging can be very helpful. For example, fresh-cut pepper can be stored in modified atmosphere bags at 5 °C for 21 days without quality loss (González-Aguilar et al. 2004). Also in case of mandarin segments, another fruit with a high metabolic activity, modified atmosphere packaging may be advantageous: Mandarin segments stored under a micro-perforated film had an internal atmosphere of 19.8% O_2 and 1.2% CO_2 and their sensory quality was positively evaluated (Del-Valle et al. 2009).

3.2.3 Hot Water Treatment

Fruit and vegetables are susceptible to microbial decay and loss of valuable compounds after harvest. Increasing awareness of chemical residues on fruit and vegetables makes it necessary to find alternative methods to reduce microbial decay especially caused by fungi. Physical treatments became more and more important in recent years to control diseases in fruit and vegetables (Usall et al. 2016). Hot water treatment as a physical treatment has an antimicrobial effect, mainly due to the temperature influence on microorganisms. Therefore, an appropriate temperature for each disease causing microorganism has to be evaluated. Hot water treatment at 52 and 56 °C for 20 s up to 2 min can reduce post-harvest decay of apples and citrus fruit caused by *Gloeosporium* and *Penicillium digitatum* (Porat et al. 2000; Trierweiler et al. 2003; Maxin et al. 2014). Furthermore, hot water treatments of only parts of vegetables like stems of broccoli can reduce the senescence and preserve the product colour (Perini et al. 2017). According to two recent reports (Trierweiler et al. 2003; Auinger et al. 2005), the basic quality parameters of apples like titratable acidity, antioxidative capacity, total phenolics, and vitamin C are not negatively influenced by hot water treatment. The impact of hot water treatment on the metabolite profile and the overall quality of fruit has, however, not yet been studied in detail. Another important field of hot water treatment are fresh-cut fruit, especially tropical fruit which are more perishable than fresh-cut apples or pears. Hot water treatment of whole mangoes before processing shows a global beneficial effect on the nutrient quality of the fresh-cut product and a better acceptability over a shelf-life of 6 days (Djioua et al. 2009). Therefore, product specific hot water treatment is a good opportunity to reduce post-harvest diseases and preserve sensory and nutritious quality of fruit and vegetables without leaving chemical residues behind (Trierweiler et al. 2003). An appropriate treatment temperature and time is important to prevent for example apples of superficial scald.

3.2.4   UV-C Treatment

Another physical non-thermal processing technology for preservation of fruit and vegetable quality and safety is the UV-C treatment. The use of UV-C light reduces microbiological deterioration and quality loss by enzymatic browning in different fruit and vegetable juices (Zhang et al. 2011; Müller et al. 2014; Riganakos et al. 2017). Not only fruit and vegetable juices can be treated with UV-C to reduce microbial spoilage and changes in nutrients: Leafy vegetables exhibited higher contents of nutritious compounds after UV-C treatment possibly because of an induced stress situation followed by a plant associated defence mechanism (Gogo et al. 2018). UV-C treatment of leafy vegetables especially in developing countries can help to provide people in these countries with healthy and safe food. Additionally, UV-C treatment requires less energy than the normal thermal treatment of fruit and vegetables which are normally used for quality preservation (Riganakos et al. 2017). This could be an advantage in developing countries where energy is often limited and expensive to provide the consumer with safe and high quality food.

3.2.5   Fermentation

Fermentation is a common method for food preservation mainly where transport and storage at low temperatures are not always guaranteed. Furthermore, it is easy to carry out in households without a lot of technology. Fermented food products are highly appreciated not only because the shelf life of products is prolonged, but also because taste, aroma, texture, and digestibility are enhanced (Holzapfel 2002). Fermentation of foods like leafy vegetables can happen spontaneously, caused by the autochthonous microbiota occurring on the raw leaves. The success of this fermentation is depending on the composition of the autochthonous microflora, especially the presence of lactic acid bacteria which are able to use sugars and produce different acids and several antimicrobial substances. In contrast, fast lowering of the pH of a fermentation approach is very important to inhibit the growth of undesirable spoilage and potentially pathogenic microorganisms like *Salmonella Enteritidis* and

Listeria monocytogenes and to obtain a safe food. To achieve this goal, the use of well-characterized lactic acid starter cultures is a big advantage in comparison to a spontaneous fermentation because they produce a certain amount of acid and reduce the pH of the fermentation batch. A fermentation approach with nightshade leaves (African indigenous vegetable) using starter cultures led to a pH below 3.5 in 24 h in comparison to a pH of about 6.5 without starter cultures (Wafula 2017). The same author also found a good preservation of vitamin B_2 and E after fermentation of cowpea leaves (African indigenous vegetable) with lactic acid bacteria *Lactobacillus plantarum* BFE 5092 and *Lb. fermentum* BFE 6620 as starter cultures.

3.3 Perspective

Quality of fruit and vegetables cannot be improved after harvest. It is only possible to try to preserve the quality produced on the field. Therefore, appropriate technologies must be available during transport and storage and even at the consumer level. Generally, low temperatures should be preferred with the exception of tropical and subtropical fruit which are sensitive to chilling injuries. In addition, storage under controlled atmosphere helps to preserve fruit and vegetable quality by slowing down the metabolism of the products. In the future, the use of optimized product specific post-harvest treatments like modified atmosphere packaging, hot water treatment, and UV-C-treatment could support the preservation of high quality fruit and vegetables during transport, storage, and the supply chain and to provide health promoting foods to the consumers. It is also important for the future to develop product specific conditions of the above mentioned post-harvest treatments to take all the possible susceptibilities of different fruit and vegetables into account to achieve the best quality and safety of the products for the consumer.

3.4 Conclusion

Quality of fruit and vegetables cannot be improved after harvest. It is only possible to preserve the quality produced on the field. Therefore, appropriate technologies must be available during transport, storage, marketing, and even at the consumer level. Generally, low temperatures should be preferred with the exception of tropical and subtropical fruit which are sensitive to chilling injuries.

In the future the use of optimized product specific post-harvest treatments like modified atmosphere packaging, hot water treatment, UV-C-treatment could support the preservation of good quality of fruit and vegetables and deliver the consumer with possibly health promoting food.

Fermentation is a common post-harvest method for food preservation mainly where transport and storage at low temperatures are not always guaranteed. Fermentation of foods like leafy vegetables can happen spontaneously, caused by the autochthonous microbiota occurring on the raw leaves. Fast lowering of the pH of a fermentation approach is very important to inhibit the growth of undesirable spoilage and potentially pathogenic microorganisms like *Salmonella Enteritidis* and *Listeria monocytogenes*. To achieve this goal, the use of well-characterized lactic acid starter cultures in comparison to a spontaneous fermentation is recommended because they produce a certain amount of acid and reduce the pH of the fermentation batch in the desired time of 24 h.

In addition to the targeted determination of basic quality parameters of fruit and vegetables after different post-harvest treatments, the untargeted metabolomics approach enables an in-depth evaluation of fruit composition and thus fruit quality.

References

Aked, J. (2002). Maintaining the post-harvest quality of fruits and vegetables. In *Fruit and vegetable processing* (pp. 119–149). Boca Raton: Woodhead Publishing.

Auinger, A., Trierweiler, B., Lücke, F. K., & Tauscher, B. (2005). Influence of hot water treatment on different quality parameters of apples during storage. *Journal of Applied Botany and Food Quality, 79*(3), 154–156.

Bekele, E. A., Ampofo-Asiama, J., Alis, R. R., Hertog, M. L., Nicolai, B. M., & Geeraerd, A. H. (2016). Dynamics of metabolic adaptation during initiation of controlled atmosphere storage of 'Jonagold' apple: Effects of storage gas concentrations and conditioning. *Postharvest Biology and Technology, 117*, 9–20.

BeMiller, J. N. (2017). Carbohydrate analysis. In S. S. Nielsen (Ed.), *Food analysis* (pp. 333–360). Cham: Springer.

Benkeblia, N. (2014). Metabolomics and postharvest sciences: Challenges and perspectives. *Acta Horticulturae, 1047*, 303–308.

Brasil, I. M., & Siddiqui, M. W. (2018). Postharvest quality of fruits and vegetables: An overview. In M. W. Siddiqui (Ed.), *Preharvest modulation of postharvest fruit and vegetable quality* (pp. 1–40). Fallik and Illic, London: Academic Press.

Brizzolara, S., Santucci, C., Tenori, L., Hertog, M., Nicolai, B., Stürz, S., Zanella, A., & Tonutti, P. (2017). A metabolomics approach to elucidate apple fruit responses to static and dynamic controlled atmosphere storage. *Postharvest Biology and Technology, 127*, 76–87.

Bunzel, M., & Schendel, R. R. (2017). Determination of (total) phenolics and antioxidant capacity in food and ingredients. In S. S. Nielsen (Ed.), *Food Analysis* (pp. 455–468). Cham: Springer.

Day, B. P. F. (2002). New modified atmosphere packaging (MAP) techniques for fresh prepared fruit and vegetables. In W. Jongen, (Ed.), *Fruit and vegetable processing: Improving quality*. Boca Raton: Woodhead Publishing.

Del-Valle, V., Hernández-Muñoz, P., Catalá, R., & Gavara, R. (2009). Optimization of an equilibrium modified atmosphere packaging (EMAP) for minimally processed mandarin segments. *Journal of Food Engineering, 91*(3), 474–481.

Djioua, T., Charles, F., Lopez-Lauri, F., Filgueiras, H., Coudret, A., Freire, M., Jr., Ducamp-Collin, M. N., & Sallanon, H. (2009). Improving the storage of minimally processed mangoes (Mangifera indica L.) by hot water treatments. *Postharvest Biology and Technology, 52*(2), 221–226.

Fallik, E., & Ilic, Z. (2018). Pre-and postharvest treatments affecting flavor quality of fruits and vegetables. In M. W. Siddiqui (Ed.), *Preharvest modulation of postharvest fruit and vegetable quality* (pp. 139–168). London: Academic Press.

Felicetti, E., Mattheis, J. P., Zhu, Y., & Fellman, J. K. (2011). Dynamics of ascorbic acid in 'Braeburn'and 'Gala' apples during on-tree development and storage in atmospheres conducive to internal browning development. *Postharvest Biology and Technology, 61*(2–3), 95–102.

Gogo, E. O., Förster, N., Dannehl, D., Frommherz, L., Trierweiler, B., Opiyo, A. M., Ulrichs, C., & Huyskens-Keil, S. (2018). Postharvest UV-C application to improve health promoting secondary plant compound pattern in vegetable amaranth. *Innovative Food Science & Emerging Technologies, 45*, 426–437.

González-Aguilar, G. A., Ayala-Zavala, J. F., Ruiz-Cruz, S., Acedo-Félix, E., & Diaz-Cinco, M. E. (2004). Effect of temperature and modified atmosphere packaging on overall quality of fresh-cut bell peppers. *LWT-Food Science and Technology, 37*(8), 817–826.

Habwe, F. O., Walingo, K. M., & Onyango, M. O. A. (2008). Food processing and preparation technologies for sustainable utilization of African indigenous vegetables for nutrition security and wealth creation in Kenya. In G. L. Robertson & J. R. Lupien (Eds.), *Using food science and technology to improve nutrition and promote national development* (pp. 2–9). Toronto: International Union of Food Science & Technology (IUFoST).

Harman, J. E., & McDonald, B. (1989). Controlled atmosphere storage of kiwifruit. Effect on fruit quality and composition. *Scientia Horticulturae, 37*(4), 303–315.

Hatoum, D., Annaratone, C., Hertog, M. L. A. T. M., Geeraerd, A. H., & Nicolai, B. M. (2014). Targeted metabolomics study of 'Braeburn' apples during long-term storage. *Postharvest Biology and Technology, 96*, 33–41.

Holzapfel, W. H. (2002). Appropriate starter culture technologies for small-scale fermentation in developing countries. *International Journal of Food Microbiology, 75*(3), 197–212.

James, C., Seignemartin, V., & James, S. J. (2009). The freezing and supercooling of garlic (Allium sativum L.). *International Journal of Refrigeration, 32*(2), 253–260.

Jom, K. N., Frank, T., & Engel, K. H. (2011). A metabolite profiling approach to follow the sprouting process of mung beans (Vigna radiata). *Metabolomics, 7*(1), 102–117.

Karadag, A., Ozcelik, B., & Saner, S. (2009). Review of methods to determine antioxidant capacities. *Food Analytical Methods, 2*(1), 41–60.

Karapanos, I. C., Chandra, M., Akoumianakis, K. A., Passam, H. C., & Alexopoulos, A. A. (2015). The ripening and quality characteristics of cherry tomato fruit in relation to the time of harvest. *Acta Horticulturae, 1079*, 495–500.

Latocha, P., Krupa, T., Jankowski, P., & Radzanowska, J. (2014). Changes in postharvest physico-chemical and sensory characteristics of hardy kiwifruit (Actinidia arguta and its hybrid) after cold storage under normal versus controlled atmosphere. *Postharvest Biology and Technology, 88*, 21–33.

Leisso, R., Buchanan, D., Lee, J., Mattheis, J., & Rudell, D. (2013). Cell wall, cell membrane, and volatile metabolism are altered by antioxidant treatment, temperature shifts, and peel necrosis during apple fruit storage. *Journal of Agricultural and Food Chemistry, 61*(6), 1373–1387.

Lopez-Sanchez, P., De Vos, R. C. H., Jonker, H. H., Mumm, R., Hall, R. D., Bialek, L., Leenman, R., Strassburg, K., Vreeken, R., Hankemeier, T., & Schumm, S. (2015). Comprehensive metabolomics to evaluate the impact of industrial processing on the phytochemical composition of vegetable purees. *Food Chemistry, 168*, 348–355.

Mack, C., Wefers, D., Schuster, P., Weinert, C. H., Egert, B., Bliedung, S., Trierweiler, B., Muhle-Goll, C., Bunzel, M., Luy, B., & Kulling, S. E. (2017). Untargeted multi-platform analysis of the metabolome and the non-starch polysaccharides of kiwifruit during postharvest ripening. *Postharvest Biology and Technology, 125*, 65–76.

Matissek, R., Fischer, M., & Steiner, G. (2018). *Lebensmittelanalytik*. Berlin: Springer Spektrum.

Mauer, L. J., & Bradley, R. L. (2017). Moisture and total solids analysis. In S. S. Nielsen (Ed.), *Food analysis* (pp. 257–286). Cham: Springer.

Maxin, P., Williams, M., & Weber, R. W. (2014). Control of fungal storage rots of apples by hot-water treatments: A Northern European perspective. *Erwerbs-Obstbau, 56*(1), 25–34.

Mitcham, B., Cantwell, M., & Kader, A. (1996). Methods for determining quality of fresh commodities. *Perishables Handling Newsletter, 85*, 1–5.

Mizrach, A., Lu, R., & Rubino, M. (2009). Gloss evaluation of curved-surface fruits and vegetables. *Food and Bioprocess Technology, 2*(3), 300–307.

Müller, A., Noack, L., Greiner, R., Stahl, M. R., & Posten, C. (2014). Effect of UV-C and UV-B treatment on polyphenol oxidase activity and shelf life of apple and grape juices. *Innovative Food Science & Emerging Technologies, 26*, 498–504.

Nardozza, S., Boldingh, H. L., Osorio, S., Höhne, M., Wohlers, M., Gleave, A. P., MacRae, E. A., Richardson, A. C., Atkinson, R. G., Sulpice, R., & Fernie, A. R. (2013). Metabolic analysis of kiwifruit (Actinidia deliciosa) berries from extreme genotypes reveals hallmarks for fruit starch metabolism. *Journal of Experimental Botany, 64*(16), 5049–5063.

Ndlela, S., Ekesi, S., Ndegwa, P. N., Ong'amo, G. O., & Mohamed, S. A. (2017). Post-harvest disinfestation of Bactrocera dorsalis (Hendel)(Diptera: Tephritidae) in mango using hot-water treatments. *Journal of Applied Entomology, 141*(10), 848–859.

Nicolai, B. M., Pedreschi, R., Geeraerd, A., Vandendriessche, T., & Hertog, M. L. A. T. M. (2010). Postharvest metabolomics. *Acta Horticulturae, 880*, 369–376.

Obenland, D., Collin, S., Sievert, J., Negm, F., & Arpaia, M. L. (2012). Influence of maturity and ripening on aroma volatiles and flavor in 'Hass' avocado. *Postharvest Biology and Technology, 71*, 41–50.

Odriozola-Serrano, I., Soliva-Fortuny, R., & Martín-Belloso, O. (2009). Influence of storage temperature on the kinetics of the changes in anthocyanins, vitamin C, and antioxidant capacity in fresh-cut strawberries stored under high-oxygen atmospheres. *Journal of Food Science, 74*(2), C184–C191.

Organisation for Economic Co-operation and Development (OECD) (2009). Guidance on objective tests to determine quality of fruits and vegetables fresh and dry and dried produce. Available from https://www.ble.de/SharedDocs/Downloads/EN/Nutrition-Food/Quality-Control/BestimmungFruechteEN.pdf?__blob=publicationFile&v=1

Perini, M. A., Sin, I. N., Jara, A. M. R., Lobato, M. E. G., Civello, P. M., & Martínez, G. A. (2017). Hot water treatments performed in the base of the broccoli stem reduce postharvest senescence of broccoli (Brassica oleracea L. Var italic) heads stored at 20 °C. *LWT-Food Science and Technology, 77*, 314–322.

Picó, Y., La Farre, M., Segarra, R., & Barceló, D. (2010). Profiling of compounds and degradation products from the postharvest treatment of pears and apples by ultra-high pressure liquid chromatography quadrupole-time-of-flight mass spectrometry. *Talanta, 81*(1–2), 281–293.

Porat, R., Daus, A., Weiss, B., Cohen, L., Fallik, E., & Droby, S. (2000). Reduction of postharvest decay in organic citrus fruit by a short hot water brushing treatment. *Postharvest Biology and Technology, 18*(2), 151–157.

Putnik, P., Kovačević, D. B., Herceg, K., Roohinejad, S., Greiner, R., Bekhit, A. E. D. A., & Levaj, B. (2017). Modelling the shelf-life of minimally-processed fresh-cut apples packaged in a modified atmosphere using food quality parameters. *Food Control, 81*, 55–64.

Raffo, A., Baiamonte, I., & Paoletti, F. (2008). Changes in antioxidants and taste-related compounds content during cold storage of fresh-cut red sweet peppers. *European Food Research and Technology, 226*(5), 1167–1174.

Riganakos, K. A., Karabagias, I. K., Gertzou, I., & Stahl, M. (2017). Comparison of UV-C and thermal treatments for the preservation of carrot juice. *Innovative Food Science & Emerging Technologies, 42*, 165–172.

Roser, S. A. T., Bernhard, B., Peter, G., Kuhnert, H., Haß, M., & Banse, M. (2013). Lebensmittelverluste bei Äpfeln. Produktions- und Vermarktungsprozesse; Vorliegende Daten zu Lebensmittelverlusten bei Äpfeln; Hochrechnung der Verluste für Deutschland. In G. Peter, K. Heike, M. Haß, M. Banse, S.A.T Roser, B. Trierweiler, & C.B. Adler (Eds.) *Einschätzung der pflanzlichen Lebensmittelverluste im Bereich der landwirtschaftlichen Urproduktion: Bericht im Auftrag des Bundesministeriums für Ernährung, Landwirtschaft und Verbraucherschutz (BMELV).*

Rudell, D. R., Mattheis, J. P., & Curry, E. A. (2008). Prestorage ultraviolet– white light irradiation alters apple peel metabolome. *Journal of Agricultural and Food Chemistry, 56*(3), 1138–1147.

Sánchez-Rangel, J. C., Benavides, J., Heredia, J. B., Cisneros-Zevallos, L., & Jacobo-Velázquez, D. A. (2013). The Folin–Ciocalteu assay revisited: Improvement of its specificity for total phenolic content determination. *Analytical Methods, 5*(21), 5990–5999.

Schirmer, H. (2001). Hinweise zur Lagerung der Apfelsorte Rubinette. *Obstbau: die Fachzeitschrift für den Obstbau-Profi, 26*(10), 539–541.

Simões, A. D., Allende, A., Tudela, J. A., Puschmann, R., & Gil, M. I. (2011). Optimum controlled atmospheres minimise respiration rate and quality losses while increase phenolic compounds of baby carrots. *LWT-Food Science and Technology, 44*(1), 277–283.

Sivakumar, D., Jiang, Y., & Yahia, E. M. (2011). Maintaining mango (Mangifera indica L.) fruit quality during the export chain. *Food Research International, 44*(5), 1254–1263.

Trierweiler, B., Schirmer, H., & Tauscher, B. (2003). Hot water treatment to control Gloeosporium disease on apples during long-term storage. *Journal of Applied Botany, 77*(5/6), 156–159.

Tyl, C., & Sadler, G. D. (2017). pH and titratable acidity. In S. S. Nielsen (Ed.), *Food analysis* (pp. 389–406). Cham: Springer.

Usall, J., Ippolito, A., Sisquella, M., & Neri, F. (2016). Physical treatments to control postharvest diseases of fresh fruits and vegetables. *Postharvest Biology and Technology, 122*, 30–40.

Wafula, E. N. (2017). Effects of postharvest-processing technologies on the safety and quality of African indigenous leafy vegetables. Unpublished Dissertation. Resource document. http://ediss.sub.uni-hamburg.de/volltexte/2017/8777/.

Wafula, E. N., Franz, C. M., Rohn, S., Huch, M., Mathara, J. M., & Trierweiler, B. (2016). Fermentation of african indigenous leafy vegetables to lower post-harvest losses, maintain quality and increase product safety. *African Journal of Horticultural Science, 9*, 1–13.

Waterhouse, A. L. (2002). Determination of total phenolics. *Current Protocols in Food Analytical Chemistry, 6*(1), I1.1.1–I1.1.8.

Winter, F., & Link, H. (2002). Lucas' Anleitung zum Obstbau (32. Auflage), Stuttgart-Hohenheim: Ulmer.

Wojciechowska, E., Weinert, C. H., Egert, B., Trierweiler, B., Schmidt-Heydt, M., Horneburg, B., Graeff-Hönninger, S., Kulling, S. E., & Geisen, R. (2014). Chlorogenic acid, a metabolite identified by untargeted metabolome analysis in resistant tomatoes, inhibits the colonization by Alternaria alternata by inhibiting alternariol biosynthesis. *European Journal of Plant Pathology, 139*(4), 735–747.

Zhang, C., Trierweiler, B., Li, W., Butz, P., Xu, Y., Rüfer, C. E., Ma, Y., & Zhao, X. (2011). Comparison of thermal, ultraviolet-c, and high pressure treatments on quality parameters of watermelon juice. *Food Chemistry, 126*(1), 254–260.

Chapter 4
Nutritional and Industrial Relevance of Particular Neotropical Pseudo-cereals

Catalina Acuña-Gutiérrez, Stefanny Campos-Boza, Andrés Hernández-Pridybailo, and Víctor M. Jiménez

4.1 Introduction

Commonly consumed cereal grains (e.g., wheat, maize, rice, barley, rye, oats, sorghum and millet), cultivated in both temperate and tropical zones, play an important role in world food security by direct consumption, but also to be included in processed foods and beverages and as animal feed. In addition to these grains, there is an increasing number of "minor grains", comprising cereals and pseudo-cereals, that have gained importance in the past years, because of the re-born of traditional consumption habits, health issues and their adaptability to particular growing conditions and, in some cases, to a shifting climate (e.g., einkorn, emmer, durum wheat, spelt, buckwheat, old world pulses, etc.) (Mir et al. 2018).

There are pseudo-cereals (cereal-like plants, in terms of utilization of their fruits and seeds that do not belong to the grass family) of Neotropical origin (because of the latitude and not necessarily of the growing conditions as some of them grow at high altitudes) that already constitute important food sources in particular regions, which are also showing a positive trend in terms of more widespread cultivation and consumption (Mir et al. 2018). In this chapter, we are going to refer particularly to the Neotropical common bean, amaranth, quinoa, chia, chan, jícaro seeds, ojoche and the Andean lupine. Some of them have been widely used for a long time and at large scale (e.g., common beans), while others are rapidly gaining importance in

C. Acuña-Gutiérrez (✉) · S. Campos-Boza · A. Hernández-Pridybailo
CIGRAS, Universidad de Costa Rica, San Pedro, Costa Rica
e-mail: catalina.acuna@ucr.ac.cr; stefanny.campos@ucr.ac.cr; andres.hernandezpridybailo@ucr.ac.cr

V. M. Jiménez
CIGRAS, Universidad de Costa Rica, San Pedro, Costa Rica

IIA, Universidad de Costa Rica, San Pedro, Costa Rica
e-mail: victor.jimenez@ucr.ac.cr

© Springer Nature Switzerland AG 2019
C. Piatti et al. (eds.), *Food Tech Transitions*,
https://doi.org/10.1007/978-3-030-21059-5_4

many countries (e.g., amaranth, quinoa and chia). There is a third group that, although has called the attention of research groups, still lacks behind in terms of commercialization and popular awareness.

4.2 Common Bean (*Phaseolus vulgaris*)

The common bean is an important crop in many developing countries because of its nutritional value. Estimations show that more than 300 million people rely on this crop for a large proportion of their protein intake, mainly in Eastern Africa and Latin America and the Caribbean. Highest consumption is registered in Africa (Burundi, Kenya and Rwanda); however, data can be underestimated (reviewed by Petry et al. 2015). Due to the extensive amount of investigations that have been conducted in terms of nutritional and nutraceutical value, key recent reviews will be highlighted briefly hereunder. In terms of their nutritional contents, common bean seeds are mainly composed by carbohydrates (56–77% dry basis – db), being starch the major one, and by proteins (around 18–28% db). An important fraction of the carbohydrates found in *P. vulgaris* grains includes insoluble fiber (ca. 20%), which consists mainly of cellulose, hemicellulose and lignin. Hayat et al. (2014) pointed out that the protein fraction contains lysine (76 mg g^{-1} protein), an amino acid that is deficient in cereal grains like maize, wheat or rice (27, 28 and 37 mg g^{-1} protein, respectively). Regarding mineral and vitamin composition, beans are superior to most cereals, and the amount present is higher when compared to that of other legumes (reviewed by Los et al. 2018) (Table 4.1). As mentioned before, common bean is not only rich in nutrients, but also has been studied in terms of its nutraceutic properties (reviewed by Suárez-Martínez et al. 2016). Functional properties of this staple food include its high contents of phenolic compounds, which act as antioxidants and have an anti-inflammatory effect (Ganesan and Xu 2017; Yang et al. 2018). Furthermore, beans contain a range of different bioactive compounds for lipid lowering (Ramírez-Jiménez et al. 2015) and α-amylase inhibitors as anti-hyperglycemic agents (Obiro et al. 2008). In addition, common bean hydrolysates and peptides have been proposed to have antimicrobial and tumor cell inhibition activity (reviewed by Luna-Vital et al. 2015). Consequently, this crop has been regarded as strategic in order to improve human health by means of reducing the risk of diabetes and overweight (Barrett and Udani 2011) and, cardiovascular (Padhi and Ramdath 2017) and digestive diseases (Awika et al. 2018; Tao et al. 2018). Therefore, common bean renders an interesting ingredient for the preparation of innovative food products.

Milling technologies to produce common bean flour have been developed and improved (reviewed by Vishwakarma et al. 2018). Bean flour has been mainly incorporated into bakery products and snacks, in many cases with demonstrated improved nutritional properties. Figueroa-González et al. (2015) used a blend of wheat and bean flour to elaborate bars and pancakes. They found higher values of protein, ash, fat and dietary fiber in the bean-supplemented products, with lower

Table 4.1 Nutritional and antinutritional compounds reported in Neotropical pseudo-cereals

Grain	Carbohydrate fraction	Protein fraction	Lipid fraction	Other compounds of interest	Antinutritional components	Reference
Common bean	55.9–77.4%	17.7–27%	0.7–2.7%	TDF (1.5–27.2%), IDF (13.9–21.5%), lysine (76 mg/g protein)	RFO (4.2%), tannins (2618.8 (mg [+]-catechin equivalents 100 g⁻¹), trypsin inhibitors (26.8 TIU mg⁻¹), lectins (8573.1 HAU mg⁻¹ protein) and phytic acid (0.4–0.6%)	De Mejia et al. (2005), Hayat et al. (2014), and Los et al. (2018)
Amaranth	48–69%[a], 3.1%[b]	13.1–20.0%	5.6–10.9%	TDF (8.8%)	Oxalates and nitrites (3–16.5 mg g⁻¹), phytic acid (0.3–0.6%), saponins (0.09–0.1%)	Rastogi and Shukla (2013) and Narwade and Pinto (2018)
Quinoa	61.2–72.6%	12.9–16.5%	5.2–9.7%	α-linolenic acid (3.8%–8.3%)	Tannins (500 mg 100 g⁻¹)	Kozioł (1992) and Valcárcel-Yamani and Lannes (2012)
					Saponins (>0.01 g 100 g⁻¹ edible portion)	
Chia	34.57 g 100 g⁻¹	25.32 g 100 g⁻¹	25.98–30.22 g 100 g⁻¹	TDF (37.50 g 100 g⁻¹), IDF (35.07 g 100 g⁻¹)	NIA	da Silva Marineli et al. (2014)
Chan	37.07–56.81%	14–22.26%	14.3–33.59%	ADF (27.5%)	Lectins (NDA)	Bird (1959), Earle and Jones (1962), Weber et al. (1991), and Aguirre et al. (2012)
Jícaro	5.5%[b]	44%	38%	Polyphenols (201 mg 100 g⁻¹ d.w.)	AANF	Corrales et al. (2017b)
				Tannins (0.16 mg 100 g⁻¹ d.w.)		
Ojoche	NIA	12.8%	NIA	Tryptophan 2.3%	NIA	Peters and Pardo-Tejeda (1982)

(continued)

Table 4.1 (continued)

Grain	Carbohydrate fraction	Protein fraction	Lipid fraction	Other compounds of interest	Antinutritional components	Reference
Andean lupine	32.9%	40%	20%	NIA	Saponins, phytic acid, tannins	Schoeneberger et al. (1982) and Carvajal-Larenas et al. (2016)

AANF absence of anti-nutritional factors
ADF acid detergent fiber
d.w. dry weight
HAU hemagglutinin activity units
IDF insoluble dietary fiber
NDA no data available
NIA no information available
RFO raffinose family oligosaccharides
TDF total dietary fiber
TIU trypsin inhibitory units
[a]Starch
[b]Sucrose

carbohydrate levels than in the commercial products. Additional examples include 50% substitution of wheat flour with that of common beans in cookies (Bonilla et al. 2017), supplementation of whole wheat bread with freeze-dried black bean seed coat extract, addition of common bean flour to rice flour and sugar to produce extruded breakfast flakes and of common bean flour into potato and tortilla chips, among others (reviewed by Mecha et al. 2018). Moreover, *P. vulgaris* flour was added into bologna sausages with the aim of reducing the amount of fat usually present (García et al. 2013). Besides finding differences in the protein and fat contents in comparison to the standard product, noteworthy was the reduced levels of residual nitrites measured after the thermal process. Moreover, the color of the product was not altered by the addition of the bean flour, but the firmness increased with higher bean flour content, a drawback for the use of the bean flour. Information available points out to the health benefits of increasing common bean consumption in the diet (Suárez-Martínez et al. 2016).

It is important to mention that common beans have some anti-nutritive compounds that include polyphenols (condensed tannins and anthocyanins), protease inhibitors, lectins and phytic acid. However, there is evidence that these components are also involved in the prevention of some chronic disorders such as cancer, heart disease and diabetes (De Mejia et al. 2005).

4.3   Amaranth (*Amaranthus* spp.)

Amaranth (*Amaranthus* spp.) was cultivated by the Mayas since about 8000 years ago and until the Spaniards settle. According to available information, indigenous inhabitants planted thousands of hectares of this grain (Rastogi and Shukla 2013).

The main component in amaranth seeds is starch (48–69% db according to the species), although it is also possible to extract oil from two species, *A. cruentus* and *A. hypochondriacus*. This oil is highly unsaturated with pleasant sensorial characteristics and is a good source of omega fatty acids. One advantage of the amaranth oil is that it can raise HDL cholesterol and lower non-HDL cholesterol in blood. In comparison with other grains, amaranth contains more protein. The essential amino acid index of amaranth proteins equals that of egg proteins and the lysine contents are high (40–50 mg g^{-1}). Regarding the mineral composition, analyses have shown the presence of K, Ca, Mg, Zn, Fe, Mn and Ni, at levels much higher than those of common cereals, so as it is the contents of vitamins C and B_2. It is also a good source of vitamin E. On the downside, amaranth contains oxalates and nitrites, known antinutritional compounds, because they inhibit the absorption of Ca and Zn. However, they can be easily removed by boiling the seeds for 5 min in water (Rastogi and Shukla 2013; Assad et al. 2017) (Table 4.1). There is evidence that the consumption of amaranth grains causes positive biological and therapeutic effects, such as protection against oxidative stress and inflammation, retardation of tumor growth and decrease of mean arterial pressure (reviewed by Algara-Suárez et al. 2016). These effects are apparently related to the high contents of squalene, flavonoids, isoprenoids and lunasin. The grain of amaranth can be eaten as breakfast cereal, incorporated into soups and in energy bars along with other grains or seeds. It can also be processed in various ways: popped, shredded and grinded to produce flour, and as an alternative to improve the nutritional value of breads, cereals, cookies and other baked goods (reviewed by Narwade and Pinto 2018), especially the protein content (de la Barca et al. 2010). Capúz and Pilamala (2015) reported that, due to its properties related to water absorption, gel formation, emulsification and protein content, amaranth flour presented good acceptability without affecting the physico-chemical and sensory characteristics of scalded sausages. An additional product obtained from ground amaranth grains is the amaranth craft beer, which according to González-Ramírez et al. (2013) does not present any alteration in the fermentation process for the production of alcohol, and provides a higher protein profile compared to a traditional commercial beer, presenting at the end a good sensorial acceptance by consumers. The nutritional aspects of this grain can contribute to balance deficient nutritional status by its inclusion into the diet as raw grain or in the preparation of processed food.

4.4   Quinoa (*Chenopodium quinoa*)

Quinoa, like amaranth, was an important crop for pre-Columbian civilizations. It is becoming more popular nowadays due to its nutraceutical properties. Main producing countries are by far Peru and Bolivia, although it is also cultivated in Ecuador, Chile, USA, China, Canada, India and some European countries. Because of the low prolamin contents, quinoa is considered a gluten-free grain. Additional beneficial properties include its high amount of α -linolenic acid (3.8–8.3%), which might counteract some degenerative diseases (Valcárcel-Yamani and Lannes 2012) (Table 4.1), and of some oxyprenilated secondary metabolites, which have been studied for their anti-cancer and anti-inflammatory properties (Fiorito et al. 2018).

There are bitter and sweet varieties of quinoa. Bitterness depends on the content of saponins in the grain (below 0.11% is considered sweet) and modifies sensory acceptance of quinoa-supplemented products and may interfere with nutrient digestion and absorption (Suárez-Estrella et al. 2018). According to Kozioł (1992) this antinutritional saponins can be removed by washing or abrasive dehulling, aiming at not exceeding 0.01 g in an edible portion of 100 g (Table 4.1). Besides utilizing actual and developing new sweet varieties, additional industrial processes to decrease bitterness are underway (Suárez-Estrella et al. 2018).

As mentioned above, quinoa flour is an option to prepare products for gluten-sensitive people. Therefore, the demand for this product in the market has increased and its price as well. For this reason, new methods have been developed to detect product adulteration with maize, soybean and wheat flour (Rodríguez et al. 2019). It is claimed that a quinoa-based milk substitute is more nutritious than equivalent milk substitutes (e.g., from rice, almonds or coconut) because of the higher protein content. This product can be used to make yogurt, being an alternative for people avoiding animal protein or who are allergic to cow's milk (Zannini et al. 2018). Nevertheless, the disadvantage of these plant-based milks is the high glycemic index resulting from the hydrolysis of the starch to reducing sugars like maltose or glucose (Jeske et al. 2018)

4.5   Chia (*Salvia hispanica*)

Chia (Lamiaceae) is a subtropical annual of Mesoamerican origin with food, cosmetic and medicinal uses in pre-Columbian times (reviewed by Bochicchio et al. 2015; Zettel and Hitzmann 2018). Chia seeds are a good source of carbohydrates (34.57 g 100 g^{-1}), protein (25.32 g 100 g^{-1}), oil (30.22 g 100 g^{-1}) and total dietary fiber (37.50 g 100 g^{-1}), mainly insoluble fiber (35.07 g 100 g^{-1}) (Table 4.1). Fatty acids available include α-linolenic acid (an omega 3 fatty acid), linoleic acid (omega 6), palmitic acid, oleic acid and stearic acid (da Silva Marineli et al. 2014). According to Nitrayová et al. (2014), chia has the highest contents of α -linolenic acid known in plants; and its low ratio to omega 3 fatty acids measured (Sargi et al. 2013;

Nitrayová et al. 2014) points out to health-beneficial effects by lowering the risk of coronary heart disease (Simopoulos 2016).

Chia seeds have recently been proposed as a potential food ingredient (it was approved as Novel Food in Europe in 2009), also because of their high antioxidant activity due to the presence of phenolic compounds such as myricetin, quercetin, kaempferol, chlorogenic acid and 3,4-dihydroxyphenylethanol-elenolic acid dialdehyde (3,4-DHPEA-EDA) (da Silva Marineli et al. 2014). Chia also helps maintaining good glycemic control and seems to promote weight loss, due to its high fiber content. For this reason, it could be used as a supplement in the treatment of overweight and obesity in people with type-2 diabetes (Vuksan et al. 2017). Chia seed uses include the consumption of whole seeds and seed flour, as an ingredient in cereal bars, biscuits, pasta, bread, snacks, and yogurt; as well seed-coat mucilage and oil can be extracted; however, drinks are still the major culinary use (Bochicchio et al. 2015; Zettel and Hitzmann 2018).

In a study carried out by Luna-Pizarro et al. (2013), whole chia flour was used to prepare pound cake. They found a cake color shift, because of the intrinsic color of the chia flour and less moisture loss during storage due to the chia high dietary fiber values mentioned above. Similar efforts have been conducted to substitute wheat flour with chia flour in pasta, and in regular and gluten-free bread because of the already-mentioned functional properties of the latter (Steffolani et al. 2014; Coelho and de las Mercedes Salas-Mellado 2015; Rodrigues Oliveira et al. 2015; Levent 2017; Sandri et al. 2017).

Because of its functional properties, several methods have been evaluated to extract chia oil with different outcomes in terms of yield and oil composition (Martínez et al. 2012). Regarding novel approaches for its use, very recently, Rojas et al. (2019) microencapsulated chia oil to enrich mayonnaise with polyunsaturated fatty acids. Microencapsulation aimed at protecting the oils against oxidative degradation. Utilization of the oil extraction by-products has been the focus of additional research. The chia seed expeller, a by-product obtained during oil extraction, was used as a source of peptides with antioxidant properties (Cotabarren et al. 2019). These peptides, smaller than 15 kDa, were capable of scavenging free radicals.

4.6 Chan (*Hyptis suaveolens*)

Chan (Lamiaceae), also known as bushmint or pignut, is an annual plant whose seeds were highly appreciated by pre-Columbian cultures, with similar uses as those of chia seeds. Nowadays, in Mexico and Central America, chan seeds are commercialized as whole grain or grounded, and are mostly used in beverages (Mapes and Basurto 2016). Early studies on chan seed chemical composition detected the presence of hemagglutinins (i.e., lectins) for the first time in a non-leguminous plant (Bird 1959). Chan seeds have gained interest as food option due to several nutritional and nutraceutical properties, and as a raw material for the food processing industry.

In terms of the nutritional composition, the protein contents of chan seeds are considered to be higher with respect to conventional grains like wheat, corn, rice, oats and barley (Weber et al. 1991). Moreover, the protein fraction contains 39% globulin, and also considerable amounts of essential amino acids such as tyrosine and phenylalanine with respect to other traditional grains like soybean, maize, rice and wheat, together with considerable amounts of Mg, P and Ca (Aguirre et al. 2012). In addition, Bachheti et al. (2015) determined that oil from chan seeds harvested in the Uttarakhand State in India was mostly composed of unsaturated fatty acids such as linoleic (omega 6) and oleic acid (omega 9) (Table 4.1). Chan seeds, as in the case of chia, present a seed-coat mucilage visible when imbibed in water. This mucilage has drawn attention for its interesting nutraceutic and food processing applications. Its carbohydrates can be divided into neutral and acidic fractions, and include D-xylose, D-mannose, D-galactose, D-glucose, L-fucose and 4-*O*-methyl-D-glucuronic acid (Gowda 1984). Subsequently, Aspinall et al. (1991) reported for the first time the presence a L-fuco-4-O-methyl-D-glucurono-D-xylan in a plant using *H. suaveolens* as a model. Some nutraceutical properties of chan seed mucilage have been described. For example, Mueller et al. (2017) determined that the neutral fraction of the chan seed mucilage polysaccharides have prebiotic activity, with a long lasting probiotic growth effect. Furthermore, Praznik et al. (2017) characterized the molecular structure of this neutral fraction of polysaccharides, and confirmed the prebiotic properties, and suggest its potential as a resource for the food and the pharmaceutical industries. Moreover, chan seeds may be interesting as raw material for foam and emulsion stabilizers in the food industry due to the structural diversity of its 11S globulin contents in the protein fraction (Bojórquez-Velázquez et al. 2016; De la Cruz-Torres et al. 2017). Rheological properties, such as its stability at temperatures between 20 and 60 °C, its fluid flow and viscoelasticity behavior in relation to the pH, its stability under NaCl presence, and stabilization after adding glucose to a 0.5% mucilage dispersion solution, also render an interesting raw product for the food industry as thickening and gelling agent (Pérez-Orozco et al. 2019). No reports on the actual direct application of chan seeds in the food industry were found; therefore, more studies in this area are promising.

4.7 Jícaro Seeds (*Crescentia alata*)

Crescentia alata (Bignoniaceae) natural distribution, from arid parts of Mexico to central Costa Rica, has expanded by means of seed dispersal by range horses and by humans interested in its edible large spherical fruits for use as utensils (Janzen 1982). Jicaro seed consumption was important in some pre-Columbian communities, and is still used to prepare "horchata", a very popular beverage in Nicaragua, Honduras and El Salvador. Structurally, these seeds are composed by a brown seed coat (32% of a total weight) and two white cotyledons (68% of a total weight) surrounded by a transparent cuticle. The cotyledons are composed mainly by proteins (44%), lipids (38%), sucrose (5.5%), fructose (0.5%), and minerals (3.5%)

especially P, K and Mg. The main antioxidants present in the jicaro seed are phenolics and tannins (Corrales et al. 2017c) (Table 4.1). Jicaro seeds also have oleic,
linoleic, palmitic, stearic and α-linolenic acids. The major amino acid present in
jícaro cotyledons is the essential amino acid leucine (7.4% of total amino acids)
(Corrales et al. 2017b). Regarding the nutritional level, besides the fatty acids profile and contents and the amino acid composition mentioned above, the low occurrence of antinutritional factors, such as anti-trypsin and -galactosides, is noteworthy
(Corrales et al. 2017b). These authors found that the roasting process to prepare
"horchata" does not affect the amount of micronutrients present, which supports
this way of consumption. However, depending on the conditions during roasting,
some quality parameters of the seeds may decrease, generating changes in color,
flavor, texture and structure. Therefore, Corrales et al. (2017a) proposed to use an
innovative process that combines roasting and tempering to promote the swelling
and mechanical peeling of jicaro seeds, without altering the white color of the coty-
ledon and its typical aroma, for valorization of the jicaro seed in new value-added
products.

4.8 Ojoche (*Brosimum alicastrum*)

Brosimum alicastrum (Moraceae) was also a highly utilized tree during pre-
Columbian times. Almost all parts of the tree can be consumed. Its leaves can be
used for forage, its seeds to prepare beverages, and food and its fruits, seeds, bark
and latex for medicinal purposes. From the seeds, it is possible to make a type of
dough that can be mixed with maize flour to prepare tortillas. Another gastronomic
use of the seeds is for the production of the Potzol drink and atole, in addition to
using them as ingredient in different meals. The seeds are an important source of
amino acids and are considered a complement to the deficiencies caused by the
corn-based diet, typical of some parts of Mexico and Central America. Seeds are
also a source of fiber, Ca, K, Fe, folic acid and vitamins A, B and C, and is also rich
in tryptophan (Peters and Pardo-Tejeda 1982) (Table 4.1).

Proximal analysis composition showed some similarities between corn and
Brosimum alicastrum starches, but also higher pH, clarity and color (Hue angle)
values of the latter (Pérez-Pacheco et al. 2014). Since the starch obtained from
ojoche has also low protein and ash contents it is considered of high purity and, for
this reason, it can be used for the production of high-fructose syrups. Results
obtained suggest that this starch is tolerant to high processing temperatures and,
therefore, can be used in food processes requiring this and for the manufacture of
biodegradable materials. These authors reported a yield of starch production of
300 g kg^{-1}. On the other hand, the raw flour and fiber residues from the starch
extraction have high contents of protein (10.81%) and of crude fiber (8.14%). These
characteristics make the ojoche flour an interesting alternative for the production of
food supplements and to promote food security by incorporating this product into
the diet (Ramírez-Sánchez et al. 2017).

4.9   Andean Lupine (*Lupinus mutabilis*)

Lupinus mutabilis (also known as Tarwi), is a domesticated legume cultivated by the ancient inhabitants of the central Andean region since pre-Inca times, although members of the genus are widely distributed and additional species were important in other ancient cultures around the Mediterranean as well (e.g., *L. albus*). *L. mutabilis* grains were found in tombs of the Nazca culture, and illustrated in the pottery of Tiahuanaco. However, after the Spanish conquest, this lupine was displaced by the introduction of other European crops (Mujica 1992).

Lupinus mutabilis grains are considerably richer in total nitrogen than those of others species of *Lupinus* (Santos et al. 1997). These authors also found that *L. mutabilis* seeds contain 210–240 mg of globulins g^{-1} of fresh weight, corresponding to 43–45% of the grain nitrogen. Moreover, more than 40% of the dry weight of Andean lupine grains is in the form of proteins, which is higher than that of soybeans, and approximately 20% of the dry weight is composed by oil, 7.5% by crude fiber and 3.2% by ash (Schoeneberger et al. 1982) (Table 4.1). Therefore, Carvajal-Larenas et al. (2016) pointed out that the addition of lupine or its derivatives (isolates, flour or protein concentrates) can be used to increase the nutrient content of different foods (e.g., adding it to a rice-based milk).

Lupine is known for its nutraceutical properties, including the capacity of reducing blood glucose levels (Zambrana et al. 2018). In addition, Muñoz et al. (2018) found that the hydrolysates from this bean improved the insulin receptor sensitivity and inhibited the hepatic gluconeogenesis.

The preparation of yogurt, based on *L. mutabilis* flour, is an important application in the food industry (Castañeda-Castañeda et al. 2008). Another use includes the replacement of wheat flour with lupine flour for the preparation of pasta. Ponce et al. (2018) reported that substitution of 25% wheat flour with lupine flour could be used to obtain a more nutritious pasta. However, flatulence, caused by the presence of raffinose, is considered a drawback for its wider consumption; while its yellow color might not be desirable in some cases (like in the making of white bread). For this reason, Güémes-Vera et al. (2008) developed a method for the detoxification of this compound by continuous water wash of the seeds, and achieved discoloration by using 1% citric acid on the flour, with satisfactory results.

Carvajal-Larenas et al. (2016) pointed out that the alkaloids present in this bean can be useful for medical purposes. Moreover, even though there are some antinutritional components present in the bean, such as saponins, phytic acid and tannins, they are found in very low quantities to be a health issue.

4.10 Conclusions

All pseudo-cereals mentioned above constitute potential sources of vitamins, minerals, and proteins. Including them as ingredients during the elaboration of traditionally consumed products (e.g., breads and cookies) is a way to incorporate them into the daily diet. By these means, products with better nutritional characteristics that could contribute to increase health of the population can be obtained. In addition, since some of the pseudo-cereals might have therapeutic effects, new products can be developed specifically for target populations with chronic and degenerative conditions. The fact that some of them are underutilized and lesser known, points out to the necessity of developing comprehensive approaches in terms of cultivation, prospection, utilization and transformation to fully explore their potential.

References

Aguirre, C., Torres, I., Mendoza-Hernández, G., Garcia-Gasca, T., & Blanco-Labra, A. (2012). Analysis of protein fractions and some minerals present in chan (*Hyptis suaveolens* L.) seeds. *Journal of Food Science, 77*(1), C15–C19.

Algara-Suárez, P., Gallegos-Martínez, J., & Reyes-Hernández, J. (2016). El amaranto y sus efectos terapéuticos. *Tlatemoani, 21*, 55–73.

Aspinall, G. O., Capek, P., Carpenter, R. C., Gowda, D. C., & Szafranek, J. (1991). A novel L-fuco-4-O-methyl-D-glucurono-D-xylan from *Hyptis suaveolens. Carbohydrate Research, 214*(1), 107–113.

Assad, R., Reshi, Z. A., Jan, S., & Rashid, I. (2017). Biology of amaranths. *The Botanical Review, 83*(4), 382–436.

Awika, J. M., Rose, D. J., & Simsek, S. (2018). Complementary effects of cereal and pulse polyphenols and dietary fiber on chronic inflammation and gut health. *Food & Function, 9*(3), 1389–1409.

Bachheti, R. K., Rai, I., Joshi, A., & Satyan, R. S. (2015). Chemical composition and antimicrobial activity of *Hyptis suaveolens* Poit. seed oil from Uttarakhand State, India. *Oriental Pharmacy and Experimental Medicine, 15*(2), 141–146.

Barrett, M. L., & Udani, J. K. (2011). A proprietary alpha-amylase inhibitor from white bean (*Phaseolus vulgaris*): A review of clinical studies on weight loss and glycemic control. *Nutrition Journal, 10*(1), 24.

Bird, G. W. G. (1959). Anti-A Hæmagglutinins from a non-leguminous plant – *Hyptis suaveolens* Poit. *Nature, 184*(4680), 109.

Bochicchio, R., Philips, T. D., Lovelli, S., Labella, R., Galgano, F., Di Marisco, A., Perniola, M., & Amato, M. (2015). Innovative crop productions for healthy food: the case of chia (*Salvia hispanica* L.). In: Vastola A (ed) *The sustainability of agro-food and natural resource systems in the Mediterranean Basin*. Springer International Publishing, Cham, (pp. 29–45).

Bojórquez-Velázquez, E., Lino-López, G. J., Huerta-Ocampo, J. A., Barrera-Pacheco, A., de la Rosa, A. P. B., Moreno, A., Mancilla-Margalli, N. A., & Osuna-Castro, J. A. (2016). Purification and biochemical characterization of 11S globulin from chan (*Hyptis suaveolens* L. Poit) seeds. *Food Chemistry, 192*, 203–211.

Bonilla, A. R., Cubero, E., & Reyes, Y. (2017). Bean (*Phaseolus vulgaris*) treatments effect on starch digestible fractions and consumer acceptability in the production of bean wheat cookies. *Journal of Food and Nutritional Disorders, 6*(3).

Capúz, N. G., & Pilamala, A. (2015). Elaboración de salchicha escaldada con sustitución parcial de harina de trigo por harina de amaranto. *Cienc E Investig, 23*, 5–10.

Carvajal-Larenas, F. E., Linnemann, A. R., Nout, M. J. R., Koziol, M., & Van Boekel, M. A. J. S. (2016). *Lupinus mutabilis*: Composition, uses, toxicology, and debittering. *Critical Reviews in Food Science and Nutrition, 56*(9), 1454–1487.

Castañeda Castañeda, B., Manrique, M., Gamarra Castillo, F., Muñoz Jáuregui, A., Ramos, E., Lizaraso Caparó, F., & Martínez, J. (2008). Probiótico elaborado en base a las semillas de Lupinus mutabilis sweet (chocho o tarwi). *Acta Médica Peruana, 25*(4), 210–215.

Coelho, M. S., & de las Mercedes Salas-Mellado, M. (2015). Effects of substituting chia (*Salvia hispanica* L.) flour or seeds for wheat flour on the quality of the bread. *LWT- Food Science and Technology, 60*(2), 729–736.

Corrales, C. V., Achir, N., Forestier, N., Lebrun, M., Maraval, I., Dornier, M., Perez, A. M., Vaillant, F., & Fliedel, G. (2017a). Innovative process combining roasting and tempering to mechanically dehull jicaro seeds (*Crescentia alata* KHB). *Journal of Food Engineering, 212*, 283–290.

Corrales, C. V., Fliedel, G., Perez, A. M., Servent, A., Prades, A., Dornier, M., Lomonte, B., & Vaillant, F. (2017b). Physicochemical characterization of jicaro seeds (*Crescentia alata* HBK): A novel protein and oleaginous seed. *Journal of Food Composition and Analysis, 56*, 84–92.

Corrales, C. V., Lebrun, M., Vaillant, F., Madec, M. N., Lortal, S., Pérez, A. M., & Fliedel, G. (2017c). Key odor and physicochemical characteristics of raw and roasted jicaro seeds (*Crescentia alata* KHB). *Food Research International, 96*, 113–120.

Cotabarren, J., Rosso, A. M., Tellechea, M., García-Pardo, J., Rivera, J. L., Obregón, W. D., & Parisi, M. G. (2019). Adding value to the chia (*Salvia hispanica* L.) expeller: Production of bioactive peptides with antioxidant properties by enzymatic hydrolysis with Papain. *Food Chemistry, 274*, 848–856.

da Silva Marineli, R., Moraes, É. A., Lenquiste, S. A., Godoy, A. T., Eberlin, M. N., & Maróstica, M. R., Jr. (2014). Chemical characterization and antioxidant potential of Chilean chia seeds and oil (*Salvia hispanica* L.). *LWT- Food Science and Technology, 59*(2), 1304–1310.

de la Barca, A. M. C., Rojas-Martínez, M. E., Islas-Rubio, A. R., & Cabrera-Chávez, F. (2010). Gluten-free breads and cookies of raw and popped amaranth flours with attractive technological and nutritional qualities. *Plant Foods for Human Nutrition, 65*(3), 241–246.

De Mejia, E. G., Valadez-Vega, M. D. C., Reynoso-Camacho, R., & Loarca-Pina, G. (2005). Tannins, trypsin inhibitors and lectin cytotoxicity in tepary (*Phaseolus acutifolius*) and common (*Phaseolus vulgaris*) beans. *Plant Foods for Human Nutrition, 60*(3), 137–145.

De la Cruz-Torres, L. F., Pérez-Martínez, J. D., Sánchez-Becerril, M., Toro-Vázquez, J. F., Mancilla-Margalli, N. A., Osuna-Castro, J. A., VillaVelázquez-Mendoza, C. I. (2017). Physicochemical and functional properties of 11S globulin from chan (*Hyptis suaveolens* L. poit) seeds. *Journal of Cereal Science, 77*, 66–72.

Earle, F. R., & Jones, Q. (1962). Analyses of seed samples from 113 plant families. *Economic Botany, 16*(4), 221–250.

Figueroa-González, J. J., Guzmán-Maldonado, S. H., & Herrera-Hernández, M. G. (2015). Atributo nutricional y nutracéutica de panqué y barritas a base de harina de frijol (*Phaseolus vulgaris* L.). *Biotecnia, 17*, 9–14.

Fiorito, S., Epifano, F., Taddeo, V. A., & Genovese, S. (2018). Recent acquisitions on oxyprenylated secondary metabolites as anti-inflammatory agents. *European Journal of Medicinal Chemistry, 153*, 116–122.

Ganesan, K., & Xu, B. (2017). Polyphenol-rich dry common beans (*Phaseolus vulgaris* L.) and their health benefits. *International Journal of Molecular Sciences, 18*(11), 2331.

García, O., Acevedo, I., & Ruiz-Ramirez, J. (2013). Efecto de adición de la harina de Phaseolus vulgaris sobre las propiedades fisicoquímicas y sensoriales de la bologna. *Gaceta de Ciencias Veterinarias, 18*(2), 47–54.

Gowda, D. C. (1984). Polysaccharide components of the seed-coat mucilage from *Hyptis suaveolens*. *Phytochemistry, 23*(2), 337–338.

Güémes-Vera, N., Peña-Bautista, R. J., Jiménez-Martínez, C., Dávila-Ortiz, G., & Calderón-Domínguez, G. (2008). Effective detoxification and decoloration of *Lupinus mutabilis* seed derivatives, and effect of these derivatives on bread quality and acceptance. *Journal of the Science of Food and Agriculture, 88*(7), 1135–1143.

González-Ramírez, J. E., de Lira, R. F., Martínez, R. C., & Salgado, J. L. M. (2013). Perspectivas de nuevos productos a base de amaranto: cerveza artesanal de amaranto. *Tlatemoani, 14*.

Hayat, I., Ahmad, A., Masud, T., Ahmed, A., & Bashir, S. (2014). Nutritional and health perspectives of beans (*Phaseolus vulgaris* L.): An overview. *Critical Reviews in Food Science and Nutrition, 54*(5), 580–592.

Janzen, D. H. (1982). Fruit traits, and seed consumption by rodents, of *Crescentia alata* (Bignoniaceae) in Santa Rosa National Park, Costa Rica. *American Journal of Botany, 69*(8), 1258–1268.

Jeske, S., Zannini, E., Lynch, K. M., Coffey, A., & Arendt, E. K. (2018). Polyol-producing lactic acid bacteria isolated from sourdough and their application to reduce sugar in a quinoa-based milk substitute. *International Journal of Food Microbiology, 286*, 31–36.

Kozioł, M. J. (1992). Chemical composition and nutritional evaluation of quinoa (*Chenopodium quinoa* Willd.). *Journal of Food Composition and Analysis, 5*(1), 35–68.

Levent, H. (2017). Effect of partial substitution of gluten-free flour mixtures with chia (*Salvia hispanica* L.) flour on quality of gluten-free noodles. *Journal of Food Science and Technology, 54*(7), 1971–1978.

Los, F. G. B., Zielinski, A. A. F., Wojeicchowski, J. P., Nogueira, A., & Demiate, I. M. (2018). Beans (*Phaseolus vulgaris* L.): Whole seeds with complex chemical composition. *Current Opinion in Food Science, 19*, 63–71.

Luis F. De la Cruz-Torres, Jaime D. Pérez-Martínez, Mayra Sánchez-Becerril, Jorge F. Toro-Vázquez, N. Alejandra Mancilla-Margalli, Juan A. Osuna-Castro, C.I. VillaVelázquez-Mendoza (2017). Physicochemical and functional properties of 11S globulin from chan (Hyptis suaveolens L. poit) seeds. Journal of Cereal Science, 77:66–72.

Luna-Vital, D. A., Mojica, L., de Mejía, E. G., Mendoza, S., & Loarca-Piña, G. (2015). Biological potential of protein hydrolysates and peptides from common bean (*Phaseolus vulgaris* L.): A review. *Food Research International, 76*, 39–50.

Mapes, C., & Basurto, F. (2016). Biodiversity and edible plants of Mexico. In R. Lira, A. Casas, & J. Blancas (Eds.), *Ethnobotany of Mexico* (pp. 83–131). New York: Springer.

Martínez, M. L., Marín, M. A., Faller, C. M. S., Revol, J., Penci, M. C., & Ribotta, P. D. (2012). Chia (*Salvia hispanica* L.) oil extraction: Study of processing parameters. *LWT- Food Science and Technology, 47*(1), 78–82.

Mecha, E., Figueira, M. E., Patto, M. C. V., & Bronze, M. (2018). Two sides of the same coin: The impact of grain legumes on human health: Common bean (*Phaseolus vulgaris* L.) as a case study. In *Legume seed nutraceutical research*. IntechOpen, London, UK (pp. 25–46).

Mir, N. A., Riar, C. S., & Singh, S. (2018). Nutritional constituents of pseudo cereals and their potential use in food systems: A review. *Trends in Food Science & Technology, 75*, 170–180.

Mueller, M., Čavarkapa, A., Unger, F. M., Viernstein, H., & Praznik, W. (2017). Prebiotic potential of neutral oligo-and polysaccharides from seed mucilage of *Hyptis suaveolens*. *Food Chemistry, 221*, 508–514.

Mujica, A. (1992). Granos y leguminosas andinas. In E. Hernández-Bermejo & J. León (Eds.), *Cultivos marginados: otra perspectiva de 1492* (pp. 129–146). Rome: Food and Agriculture Organization of the United Nations.

Muñoz, E. B., Luna-Vital, D. A., Fornasini, M., Baldeón, M. E., & de Mejia, E. G. (2018). Gamma-conglutin peptides from Andean lupin legume (*Lupinus mutabilis* Sweet) enhanced glucose uptake and reduced gluconeogenesis in vitro. *Journal of Functional Foods, 45*, 339–347.

Narwade, S., & Pinto, S. (2018). Amaranth – A functional food. *Concepts Dairy & Veterinary Science, 1*, 72–77.

Nitrayová, S., Brestenský, M., Heger, J., Patráš, P., Rafay, J., & Sirotkin, A. (2014). Amino acids and fatty acids profile of chia (*Salvia hispanica* L.) and flax (*Linum usitatissimum* L.) seed. *Potravinarstvo Scientific Journal for Food Industry, 8*, 72–76.

Obiro, W. C., Zhang, T., & Jiang, B. (2008). The nutraceutical role of the *Phaseolus vulgaris* α-amylase inhibitor. *British Journal of Nutrition, 100*(1), 1–12.

Padhi, E. M., & Ramdath, D. D. (2017). A review of the relationship between pulse consumption and reduction of cardiovascular disease risk factors. *Journal of Functional Foods, 38*, 635–643.

Pérez-Orozco, J. P., Sánchez-Herrera, L. M., & Ortiz-Basurto, R. I. (2019). Effect of concentration, temperature, pH, co-solutes on the rheological properties of *Hyptis suaveolens* L. mucilage dispersions. *Food Hydrocolloids, 87*, 297–306.

Peters, C. M., & Pardo-Tejeda, E. (1982). *Brosimum alicastrum* (Moraceae): Uses and potential in Mexico. *Economic Botany, 36*(2), 166–175.

Petry, N., Boy, E., Wirth, J., & Hurrell, R. (2015). The potential of the common bean (*Phaseolus vulgaris*) as a vehicle for iron biofortification. *Nutrients, 7*(2), 1144–1173.

Pizarro, P. L., Almeida, E. L., Sammán, N. C., & Chang, Y. K. (2013). Evaluation of whole chia (*Salvia hispanica* L.) flour and hydrogenated vegetable fat in pound cake. *LWT- Food Science and Technology, 54*(1), 73–79.

Ponce, M., Navarrete, D., & Vernaza, M. G. (2018). Sustitución Parcial de Harina de Trigo por Harina de Lupino (*Lupinus mutabilis* Sweet) en la Producción de Pasta Larga. *Información tecnológica, 29*(2), 195–204.

Praznik, W., Čavarkapa, A., Unger, F. M., Loeppert, R., Holzer, W., Viernstein, H., & Mueller, M. (2017). Molecular dimensions and structural features of neutral polysaccharides from the seed mucilage of *Hyptis suaveolens* L. *Food Chemistry, 221*, 1997–2004.

Pérez-Pacheco, E., Moo-Huchin, R. J. Estrada-León, R. J., Ortiz-Fernández, A., May-Hernández, L. H., Ríos-Soberanis, C. R., Betancur-Ancona, D. (2014). Isolation and characterization of starch obtained from *Brosimum alicastrum* Swarts Seeds. *Carbohydrate Polymers 101*:920–927.

Ramírez-Jiménez, A. K., Reynoso-Camacho, R., Tejero, M. E., León-Galván, F., & Loarca-Pina, G. (2015). Potential role of bioactive compounds of *Phaseolus vulgaris* L. on lipid-lowering mechanisms. *Food Research International, 76*, 92–104.

Ramírez-Sánchez, S., Ibáñez-Vázquez, D., Gutiérrez-Peña, M., Ortega-Fuentes, M. S., García-Ponce, L. L., & Larqué-Saavedra, A. (2017). El Ramón (*Brosimum alicastrum* Swartz) una alternativa para la seguridad alimentaria en México. *Agroproductividad, 10*(1), 80–83.

Rastogi, A., & Shukla, S. (2013). Amaranth: A new millennium crop of nutraceutical values. *Critical Reviews in Food Science and Nutrition, 53*(2), 109–125.

Rodrigues Oliveira, M., Ercolani Novack, M., Pires Santos, C., Kubota, E., & Severo da Rosa, C. (2015). Evaluation of replacing wheat flour with chia flour (*Salvia hispanica* L.) in pasta. *Semina: Ciências Agrárias, 36*(4), 2545–2553.

Rodríguez, S. D., Rolandelli, G., & Buera, M. P. (2019). Detection of quinoa flour adulteration by means of FT-MIR spectroscopy combined with chemometric methods. *Food Chemistry, 274*, 392–401.

Rojas, V. M., Marconi, L. F. D. C. B., Guimarães-Inácio, A., Leimann, F. V., Tanamati, A., Gozzo, Â. M., Fuchs, R. H. B., Barreiro, M. F., Barros, L., Ferreira, I. C., & Tanamati, A. A. C. (2019). Formulation of mayonnaises containing PUFAs by the addition of microencapsulated chia seeds, pumpkin seeds and baru oils. *Food Chemistry, 274*, 220–227.

Sandri, L. T., Santos, F. G., Fratelli, C., & Capriles, V. D. (2017). Development of gluten-free bread formulations containing whole chia flour with acceptable sensory properties. *Food Science & Nutrition, 5*(5), 1021–1028.

Santos, C. N., Ferreira, R. B., & Teixeira, A. R. (1997). Seed proteins of *Lupinus mutabilis*. *Journal of Agricultural and Food Chemistry, 45*(10), 3821–3825.

Sargi, S. C., Silva, B. C., Santos, H. M. C., Montanher, P. F., Boeing, J. S., Júnior, S., Oliveira, O., Souza, N. E., & Visentainer, J. V. (2013). Antioxidant capacity and chemical composition in seeds rich in omega-3: Chia, flax, and perilla. *Food Science and Technology, 33*(3), 541–548.

Schoeneberger, H., Gross, R., Cremer, H. D., & Elmadfa, I. (1982). Composition and protein quality of *Lupinus mutabilis*. *The Journal of Nutrition, 112*(1), 70–76.

Simopoulos, A. P. (2016). Evolutionary aspects of the dietary omega-6/omega-3 fatty acid ratio: Medical implications. In A. Alvergne, C. Jenkinson, & C. Faurie (Eds.), *Evolutionary thinking in medicine* (pp. 119–134). Cham: Springer.

Steffolani, E., De la Hera, E., Pérez, G., & Gómez, M. (2014). Effect of chia (Salvia hispanica L) addition on the quality of gluten-free bread. *Journal of Food Quality, 37*(5), 309–317.

Suárez-Estrella, D., Torri, L., Pagani, M. A., & Marti, A. (2018). Quinoa bitterness: Causes and solutions for improving product acceptability. *Journal of the Science of Food and Agriculture, 98*(11), 4033–4041.

Suárez-Martínez, S. E., Ferriz-Martínez, R. A., Campos-Vega, R., Elton-Puente, J. E., de la Torre Carbot, K., & García-Gasca, T. (2016). Bean seeds: Leading nutraceutical source for human health. *CyTA Journal of Food, 14*(1), 131–137.

Tao, J., Li, Y., Li, S., & Li, H. B. (2018). Plant foods for the prevention and management of colon cancer. *Journal of Functional Foods, 42*, 95–110.

Valcárcel-Yamani, B., & Lannes, S. D. S. (2012). Applications of quinoa (*Chenopodium quinoa* Willd.) and amaranth (Amaranthus spp.) and their influence in the nutritional value of cereal based foods. *Food and Public Health, 2*(6), 265–275.

Vishwakarma, R. K., Shivhare, U. S., Gupta, R. K., Yadav, D. N., Jaiswal, A., & Prasad, P. (2018). Status of pulse milling processes and technologies: A review. *Critical Reviews in Food Science and Nutrition, 58*(10), 1615–1628.

Vuksan, V., Jenkins, A. L., Brissette, C., Choleva, L., Jovanovski, E., Gibbs, A. L., Bazinet, R. P., Au-Yeung, F., Zurbau, A., Ho, H. V. T., & Duvnjak, L. (2017). Salba-chia (*Salvia hispanica* L.) in the treatment of overweight and obese patients with type 2 diabetes: A double-blind randomized controlled trial. *Nutrition, Metabolism, and Cardiovascular Diseases, 27*(2), 138–146.

Weber, C. W., Gentry, H. S., Kohlhepp, E. A., & McCrohan, P. R. (1991). The nutritional and chemical evaluation of chia seeds. *Ecology of Food and Nutrition, 26*(2), 119–125.

Yang, Q. Q., Gan, R. Y., Ge, Y. Y., Zhang, D., & Corke, H. (2018). Polyphenols in common beans (*Phaseolus vulgaris* L.): Chemistry, analysis, and factors affecting composition. *Comprehensive Reviews in Food Science and Food Safety, 17*(6), 1518–1539.

Zambrana, S., Lundqvist, L., Mamani, O., Catrina, S. B., Gonzales, E., & Östenson, C. G. (2018). *Lupinus mutabilis* extract exerts an anti-diabetic effect by improving insulin release in type 2 diabetic Goto-Kakizaki rats. *Nutrients, 10*(7), 933.

Zannini, E., Jeske, S., Lynch, K. M., & Arendt, E. K. (2018). Development of novel quinoa-based yoghurt fermented with dextran producer Weissella cibaria MG1. *International Journal of Food Microbiology, 268*, 19–26.

Zettel, V., & Hitzmann, B. (2018). Applications of chia (*Salvia hispanica* L.) in food products. *Trends in Food Science & Technology, 80*, 43–50.

Chapter 5
The Demand for Superfoods: Consumers' Desire, Production Viability and Bio-intelligent Transition

Simone Graeff-Hönninger and Forough Khajehei

5.1 Introduction

Food products are a fundamental part of people's life, providing nutrition, health, and well-being. However, evolving consumer demands and changing lifestyles have provoked some kind of revolution of the food industry in recent years. Years and centuries ago sugary drinks and snacks were popular items as mankind has always consumed collected fruits and seeds associated with sweetness. Then, for decades consumers were urged to banish fat from their diets whenever possible. Fat in food products was classified in good fats, which included monounsaturated and polyunsaturated fats, bad ones which comprised industrial-made trans fats and saturated fats which fell somewhere in the middle (Mattson and Grundy 1985). In consequence, the food industry focused on creating low-fat or even no-fat food products, using fat-substitutes or other replacements and fillers. However, the shift did not make consumers healthier, probably as they cut back on both healthy as well as harmful fats and most medicals now no longer insist that fat is the one and only culprit.

As society is facing an increasing global problem of rising obesity rates and obesity-related diseases (Wyatt et al. 2006), now a shift in food product development is occurring to lowering the sugar contents. Over the last decades, sugar was added to nearly all processed foods, thus limiting consumer choice and selection possibilities for products with either less or reduced sugar amounts (Vio and Uauy 2007). As consumer awareness of possible impacts of nutrition on their health and

S. Graeff-Hönninger (✉) · F. Khajehei
Institute of Crop Science, University of Hohenheim, Stuttgart, Germany
e-mail: graeff@uni-hohenheim.de; f.khajehei@uni-hohenheim.de

© Springer Nature Switzerland AG 2019
C. Piatti et al. (eds.), *Food Tech Transitions*,
https://doi.org/10.1007/978-3-030-21059-5_5

overall daily welling has increased, food producers are forced to lower the amount of sugar added to foods and fat is no longer observed as the "one and only bad guy" in food products.

Fat and sugar are just two examples of transitions that have occurred in the food area over the last decades, forcing the food industry to come up with new concepts to meet the ever changing consumer demand. With the increase of life-expectancy and changing eating habits in the 1980s consumers now demand overall healthy products, which are advertised as 'all natural', 'free-from' and 'no added'. The demand for healthier foods has been driven by consumer's changing preferences and a greater cultural awareness of nutrition (Falguera et al. 2012). The food industry has picked up this trend and either reformulated or changed the marketing of their products. At the moment, the focus for the development of new food products is on issues of sustainable handling of food welfare and the resulting potential health consequences (obesity, diabetes, etc.) as well as changes in dietary habits, such as vegetarian, vegan, gluten-free, or low-carb to name the most common. Demographic change and increasing individualization at all levels of society require a fundamental reorientation of supply and product concepts. The food industry is called upon to actively and proactively confront the associated social demands for a needs-based supply of energy, nutrients and information. The food industry now faces the challenge of satisfying a variety of nutritional needs and eating habits by diversifying its product range.

Within the last few years, there has been a soaring interest in so-called 'superfoods' among consumers in western countries. Superfoods gained attraction due to their possible health benefits as they may contain bioactive compounds that have a relevance for either an improved state of health and well-being and/or a reduction of risk of diseases (Brunso et al. 2002). Although no official definition exists to what constitutes a 'superfood', the term is usually applied to food products that contain high amounts of nutrients (e.g. antioxidants, vitamins, minerals). In 2015, the growing popularity of superfoods has resulted in a 36% increase of newly launched products that are comprised of various superfood, supergrain, superfruit, and other similar labels (Mintel 2016). In addition, the superfood boom is driven by growing food incompatibilities and above all by a new, critical consumer type, which has a widespread mistrust of industrially manufactured products. The trends of individualization in nutrition and health form the basis for the growing demand of superfoods, especially if foods products could pose a risk to health.

Besides their potential of meeting consumer's desire for healthier, less refined, more or less also raw and clean-label food products, superfoods offer new flavors and texture components. The food industry can take advantage of this situation using these new, innovative raw materials in combination with the development of new formulations to create food products that are individually tailored to different nutritional trends and consumer demands.

5.2 Determinants of Food Choice and Consumer Behaviour

Consumers' rising awareness of the interrelation between nutrition and health are top of the list of priorities when creating and marketing new food products. Consumers' perception of the healthiness of food is influenced by its type, processing, origin, production system, which could be either organic or conventional, conservation method, packaging, and use of additives (Bech-Larsen and Grunert 2003). Additionally, diet trends like "all-raw", "free-from", "vegan" and eating habits influence the consumption of superfoods, which is often also an expression of social distinction. Consumers of superfoods are quite often found among groups of higher educational standards (Groeniger et al. 2017). Rützler and Reiter (2015) posit that food trends "display desires and attitudes to life". They are seen as "searches for solutions to problems that society is constantly facing". Kaur and Singh (2017) identified in their review four categories that affect consumers' behavior and choice of functional food and defined them as (1) personal factors, (2) psychological factors, (3) cultural and social factors, and (4) factors related to the food product. The authors point out the relevance and influence of these factors on food choice, wherefore the reader is directed to this review for further details.

Relating the factors Kaur and Singh (2017) identified to the choice of superfoods, it has to be stated that superfoods are considered as something special as they come along with creative, fancy, ancient names. Most of them are exotic, emerge from different unexplored, fascinating countries and regions, come along with mystic stories of former cultures and, after all, are rather expensive. Hence, superfoods are perceived as no ordinary food and are experienced to make out the little difference between standard eating habits. Nevertheless, the most mentioned reason for the consumption of superfoods under different consumer groups is the desire to stay healthy. A healthy diet promises a strong, healthy body and immune system. A healthy immune system is better prepared to compete with negative health effects (EUFIC 2013). Another often given motivation for the consumption of superfoods is the loss of weight (Fitschen 2015). Other motivations are based on cosmetic claims such as glowing skin and healthy looking hair (Fitschen 2015). Bugge (2015) claims that "the immense preoccupation with healthy eating can be seen as a quest for identity, spirituality and control". According to Bugge (ibid.) "superfoods are an example that fits right into an identity quest as they reflect the diets of post-secular societies".

5.3 Superfoods: Will They Stay or Will They Go?

Nowadays, superfoods are an inherent part of product lines in supermarkets. Especially in the industrialized countries, the demand for healthy food and the associated consumer interest in health-promoting ingredients in foods has risen sharply.

The question will be if the introduced superfoods will be only a "hype" and disappear shortly after they have entered the market or if they will remain and fulfill their promises and meet consumers desire for a healthy nutrition.

5.3.1 Vitamin Rich Superfoods: The Case of Moringa oleifera

Moringa oleifera Lam. is a fast-growing small tree native to the sub-Himalayan tracts of Northern India (Ayerza 2011). Moringa plants were used in traditional medicine for centuries, given the name of "the miracle tree". It has been used in India for food, feed, and medicines (Acosta 1578). The leaves are considered to be rich in protein and contain a variety of vitamins and minerals. In addition, moringa seeds contain 35–45% oil, which is used in cosmetic products. The oil is characterized by almost total natural absence of color and odor and high oleic acid concentration (>73%). The low content of polyunsaturated fatty acids (<1%) results in a high oxidative stability of the oil (Lalas and Tsaknis 2002; Kleiman et al. 2008). Because of its high nutrient concentration, moringa is considered as a supplement indicating anti-inflammatory, cardio- protective, anti-asthmatic, antibiotic, and anti-diabetic properties. While there are some research studies available substantiating the use of moringa, there are some drawbacks. Its taste is often rated as grassy, and it is not as exceptional of a nutrient source as it is often claimed to be. Especially in the developed world there are other sources of these nutrients, which do not go along with a grassy aftertaste. As the tree is so economical, supplements are cheap and easy to produce. These low overhead costs make it ideal for companies to mass produce moringa wherefore interest in it decreased, as it is no longer exposing the glory of something special and rare.

5.3.2 Gluten Free Superfoods: The Case of Quinoa

Quinoa (*Chenopodium quinoa* Willd.) belongs to the family of the Amaranthaceae, is an herbaceous seed crop, whose origins lie in the Andean region of Bolivia and Peru. Its natural spatial distribution ranges from Colombia to Chile. As the nutritional food quality of quinoa grains is considered quite high and quinoa can possibly be grown anywhere as long as the day length requirements are met, the interest in quinoa has risen enormously in recent years. Past breeding efforts created cultivars with a range of photoperiodic responses (Ruiz et al. 2014). Current breeding efforts mainly target cultivars containing little or no saponins (anti-nutritional compounds causing a bitter taste) and the improvement of traits related to grain yield like grain size, and harvest index (Zurita-Silva et al. 2014). Quinoa is considered to play an important role in eradicating hunger, malnutrition and poverty, which explains why it was declared to be the perfect food for humanity and the FAO declared the year 2013 to "The International Year of Quinoa".

Quinoa exhibits favorable nutritional properties, including high contents of protein, minerals and flavonoids. The crop is recognized as a good source of bioactive compounds, including dietary fiber, carotenoids, phytosterols, squalene, steroids and phenolic compounds (Lutz and Bascunan-Godoy 2017). Moreover, it is a good source of thiamine, folic acid and vitamin C (Pulvento et al. 2010). Other components of the grain, such as saponins and small starch granules can be of interest for industrial application. Quinoa also qualifies as gluten-free seeds suitable for consumption by people suffering from coeliac disease. It is now also used in the preparation of emulsion-type products, as milk substitute or in malted beverages. In baked foods it can be used as fat/cream substitute to enhance the quality (Delatorre-Herrera 2003; Bazile 2014; Bazile et al. 2016). Due to its high content of vitamin E and phenolic compounds and the resulting antioxidant properties and antimicrobial activity it can also be applied to food preservation (Vega-Galvez et al. 2010; Martinez et al. 2009). Quinoa products offered by European producers are mostly based on whole seeds or flour and aim to provide benefits related to its high omega-3 content and allergenic-friendly properties, respectively. Between 2014 and 2015, the percentage of food and drink products containing quinoa rose by 27%. The reintroduction of quinoa as ancient Andean grain into modern diets is often related to the increasing health problems like obesity, cardiovascular diseases, diabetes and cancer stimulating consumer's desire of healthy foods. Quinoa represents an excellent example that can contribute to a healthy diet and supplies good quality protein to support human health (Lutz and Bascunan-Godoy, 2017). Moreover, it can serve as good opportunity to strengthen regional agriculture and to diversify agricultural markets.

5.3.3 Polyunsaturated Fatty Acids: The Case of Chia

Chia (*Salvia hispanica* L.) is an annual herbaceous crop of the Lamiaceae family, native to southern Mexico and northern Guatemala (Tavares et al. 2018). In pre-Columbian times, it was considered as staple food next to maize, beans, and amaranth. Moreover, various medical uses of the plant have been reported, reflecting the early knowledge about its beneficial health effects (Cahill 2003). Within the course of the Spanish colonisation, the cultivation of chia dropped considerably and the traditional crop almost fell into oblivion. Over the past 20 years, chia seeds have re-emerged as functional "superfood" for human and animal nutrition with remarkable nutritive characteristics. The seeds contain up to 35% oil with an exceptional high share in omega-3 fatty acids (~60%; mainly α-Linolenic acid) and a low ω6:ω3 ratio, wherefore it is particularly suited as supplement for a vegetarian or vegan diet. Similar to flax, chia is supposed to have considerable health benefits to humans (Cahill 2003). Its antioxidant activity is attributed to flavonol glycosides, cholorogenic acid and caffeic acid. It was found to be higher than of many cereals (Vazquez-Ovando et al. 2009). The seed coat of chia is rich in fibers, and produces a mucilaginous gel when soaked in water. Due to their composition and properties chia protein and mucilage can be considered as raw materials for the preparation of edible films. These films display the ability to improve the overall quality of many fresh foods products, extend their shelf-life, as well as the nutritional characteristics. Furthermore, high contents of dietary fiber, high quality proteins and antioxidants turn chia seeds into a wholesome food commodity with health promoting effects especially linked to cardiovascular diseases (Coelho and de las Mercedes Salas-Mellado 2014) cancers and diabetes.

Currently, chia is mostly imported into Western countries from Argentina, Bolivia, Ecuador and Peru. Commercial chia seed yields in its country of origin generally lie between 500 and 600 kg ha^{-1} (Coates and Ayerza 1996). Yield can be substantially increased to 2500 kg ha^{-1} when irrigation and nitrogen fertilizer are applied. Chia and chia oil are used in human food products, as animal feed, drying oil in paints, and ingredients in cosmetics (Athar and Nasir 2005). Chia leaf oil may be useful in flavorings or fragrances and possibly be used as a pesticide, since white flies and other insects seem to avoid the plant (Ahmed et al. 1994). In its countries of origin, the seeds are mainly used for the preparation of traditional drinks, called "Chia fresca" or "Agua de Chia", which constitute mixtures of seeds and water with e.g. lime juice or honey as flavour enhancing ingredients. In other regions, especially North America and Australia, a broad range of different chia products (bars, pastry, cereal, chips, baby food, smoothies) is available.

Over the past two years, the percentage of food and drink products containing chia seeds has risen by 70%. Single components of chia seeds can be used for the production of functional and easy-to-use ("convenient") food products (see Fig. 5.1), while the full potential is not exploited yet. Currently only chia flour and chia oil are utilized as processed goods from chia seeds. The use of the strong hydrophilic chia

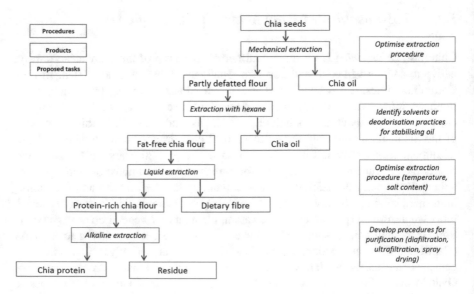

Fig. 5.1 Products and processes associated to the fractionation of chia seeds

seeds as vegetable binding agent in different (vegetarian/vegan) food products offers a special option for the food product development as it can be used for the thickening of sauces, soups, deserts, milk products, etc. In this context, the swellability of chia seeds is of major importance and is determined by the characteristics of the raw material (e.g. grain size distribution) and the preparation of the raw material (e.g. milling) offering new possibilities for the food industry explaining also the steep import increase (53% since 2013) to 18,7 T in 2017. In this context chia is no longer considered as a hype, it is recognized as superfood that will stay.

5.3.4 Alternative Sweeteners and Sugar Replacers: The Case of Yacon

The tuber fruit species yacon (*Smallanthus sonchifolius*), which is native to South America, is currently enjoying an increasing popularity in Europe due to its prebiotic advantages and benefits that are related to its high content of fructooligosaccharides and inulin, as well as phenolic compounds (Grau and Rea 1997; Ojansivu et al. 2011). Yacon tubers contain a high proportion of potentially indigestible oligosaccharides, i.e. so-called fructooligosaccharides (FOS) (up to 67% in the TS) (Ohyama et al. 1990; Hermann et al. 1999, Yun et al. 2010), which are essentially kestose (GF2), nystose (GF3), and 1-Fructofuranosyl nystose (GF4) (Roberfroid et al. 2010, Choque Delgado et al. 2010). Due to its potential high tuber yield (30–50 t), comparatively high quantities of FOS can be produced on a hectare basis. In addition, yacon has the highest fructan content (on fresh matter basis) compared to other FOS sources, such as Jerusalem artichoke or chicory (Pedreschi et al. 2003) (Voragen 1998; NRC 1989). Yacon tubers have been used as natural sweeteners and syrups for digestive problems, particularly for balancing the intestinal microbiota (Rolim 2015; Respondek et al. 2013). Especially for people suffering from diabetes and obesity, yacon products represent an alternative due to their unique sugar composition. Insulin production in the body is not stimulated, wherefore yacon has a glycemic index of 1 (Cisneros-Zevallos et al. 2002). In addition, the consumption of yacon or the contained FOS is particularly beneficial for the human digestive tract. The slightly sweet-tasting FOS cannot be degraded directly (da Silva et al. 2002), act as a prebiotic and increase the growth of beneficial intestinal bacteria (e.g. bifidobacteria and lactobacillus). This makes yacon to an ideal raw material for the production of sugar-reduced foods. The Andean population further attributes anti-diabetic properties to yacon leaves and uses them in the preparation of teas as part of low calorie diets (Campos et al. 2012). Yacon tea is also prescribed to patients who suffer from digestive and kidney diseases (Pineiro et al. 2008). In addition to prebiotics, yacon contains flavonoids, phenolic acids and tryptophan, which display antioxidant, anti-inflammatory, antimicrobial and anticancer activities. The phenolic compounds protect biomolecules, such as DNA, lipids and proteins, against damage caused by free radicals

(Jiménez and Sammán 2014). Due to the high demand for sugar-reduced food, the use of yacon represents a great opportunity for both innovation and adding value in food products (Caetano et al. 2016). The FOS stored in the tubers are not subject to the Maillard reaction. With regard to the required stability, they are stable at pH values>3 and temperatures up to 140 °C. This would preserve FOS in the most food thermal processes (Santana and Cardoso 2008), which is highly relevant to the development of new yacon-based formulations. Yacon tubers can be used either raw, cooked as soup, roasted or dried (Vilhena et al. 2000), or processed (Genta et al. 2005) as jam (Prati et al. 2009), syrup (Manrique et al. 2005), vinegar (Hondo et al. 2000), flour (Moscatto et al. 2006; Rosa et al. 2009), crisps (dried, sliced yacon tubers), and juice (Santana and Cardoso 2008). In Japan, breads, fermented drinks, freeze-dried powder, juice and other products are created from the flours (Santana and Cardoso 2008). Numerous studies have already shown that the use of inulin- and oligofructose-containing flours in various food products meet with consumer acceptance, as the rheological properties were rated as particularly good. The products obtained are easy to chew, have an airy texture, good taste and are easy to digest (Leitão et al. 1984; Moscatto et al. 2006). The same could be shown for the use of yacon syrup as a sugar substitute in corresponding food or bakery products (Maldonado and Singh 2008).

Overall, there is great potential for food products generated out of yacon. In this sense, the National Research Council ranked yacon as a promising raw material based on its high fructans content (De Almeida et al. 2015) and as a source for the development of natural sweeteners and syrups for people with digestive problems. The food industry has to depict the given benefits and test yacon as sugar replacer finding new formulations and ways to substitute in this context also the functions of common sugars as bulking agent or filler to increase the volume and viscosity of a food product. Nevertheless, besides the seemingly benefits of the above given examples of superfoods it has to be stated that nutritious and relatively inexpensive fresh foods — like kale, tomatoes, blueberries, broccoli etc. can be often produced locally and sustainably. These foods are stuffed with beneficial phytochemicals, are high in protein, loaded with polyunsaturated fatty acids and if eaten in combination, provide the necessary amounts of vitamins and minerals. They just sound less exotic and the marketing stories build around them are less mystic and fascinating, making them less appealing.

5.4 Major Food System Changes

In the last decades, the landscape of the food industry has continuously been progressing and changing with the need to address the challenges of food production under a still growing population, changing climate, and new consumer demands. Many established companies and food retailers have acquired smaller companies or start-ups specializing in foods labeled as natural, healthy, and organic to expand their product lines to niche and specialty markets. Especially the start-up community has invested in high-tech food ventures with mission statements framed by social and environmental welfare. Other changes include the adoption of different labels like "clean", "raw", "pure" to attract consumers by a more transparent formulation of their products. As consumers affect the marketplace and the success of a product by their purchasing choices and concerns, the food industry keeps their eyes open to emerging and rising as well as changing consumer demand and interests. Food technology will continue to evolve in the future, creating new products to meet the changing consumer desire, taking other restrictions associated with agricultural production, climate change and limited land area food can be produced on into account. However, the direction that it takes will be strongly tied to the complex interplay of market and consumer demand. While the technical development of hyperpalatable foods could lead to extreme food addiction on a global scale (Pelchat et al., 2004) the industry could also shift towards the large-scale adoption of non-thermal processing technologies that minimally affect the nutritional properties of a food product giving low-income group the chance to purchase low cost, nutrient-rich foods with enhanced shelf-lives. While it is not clear which shift and direction the food industry will take in the future, scientists can push the scientific knowledge and help to create a decision matrix for each potential direction. One needs to keep in mind that the advancement of food technology has many possibilities that directly impact the health of populations in the long term.

As food products change with changing consumer desires and available food technology, it gets obvious that the food production system will change to meet new needs. It's already under natural economic pressures, struggling to keep pace with consumption needs. To battle scarcity, react to climate changes, and meet the consumption needs of a growing urban population, the food production system will have to be rethought. Food production will have to be reinvented, and even stages of the food production system will shift. Product design might become more creative using new materials like superfoods to create new products. All of these changes point to a different food production system in the near future that evolves quickly to meet consumers' needs. New processes have to be developed and combined with suitable raw material to finally create overall healthy food. At the technological level, the food industry is currently undergoing an optimization cycle. The associated successes with regard to the economical use of resources, the appropriate utilization of process by-products and the increased product quality are, however, limited by the current, largely traditional product portfolio. Dairy, meat and cereal products such as cheese, sausage and bakery products originate in their present form and composition of human developmental history. Characteristic of many of these products is a high energy density with limited amount of nutrients. This significantly increases the risk of an over-energy diet due to changes in the age structure as well as the world of work and consumption. So far, food processing companies have failed to provide systematically forward-looking solutions to the question of how to redesign the current tension between the challenges and opportunities associated with the trends of digitization and demographic change. As shown by a number of recent examples, businesses are currently pursuing the strategy, under some massive public and political pressure, of reacting to criticism of their products and processes rather than actively shaping the debate on the food of the future and advance. However, an interdisciplinary, scientific examination of the nutritional and culinary properties of processed foods is required in order to remain capable of innovation and action and thus to remain competitive under the public observation that is still to be expected.

5.5 Bio-intelligent Transition

New raw materials emerging from superfoods might play a role in meeting future challenges of the food sector and offer some options, which merit particular attention. Based on sound production systems they might offer options for increasing agricultural productivity whilst adapting to the effects of climate change. If locally produced, they can also reduce emissions associated with transport of food products over long distances. In addition, the diversification of cropping systems by including superfood crops will offer means of reversing continued declines in farmland biodiversity. Based on new food technologies they might also help to reduce food wastage. The options for using wastes and by-products will help to meet further needs in a sustainable way. With the help of biological transformation in the food sector, new groundbreaking innovations will be possible creating new, bio-intelligent values in the areas of health, nutrition, consumption, and overall well-being.

5.6 Conclusion

To make an effective change, a concerted effort between the agricultural production and the food technology sector including the consumer is needed to finally move forward on the challenges of global health and nutrition. Some food companies have already become more transparent about the ingredients and production practices used to create modern foods. Creating a unified communication may be an important step in the future to retain consumer trust, while reducing the negative health impacts of certain low-cost food products on populations of lower socio-economic status as well as in the promotion of beneficial health aspects. Regardless, emerging bio-intelligent combinations of suitable raw materials and inspiring food technologies may guide and shape the food sector of the future. Food technology and scientists will continue to play an important role in shaping the resulting food products.

References

Acosta, C. (1578). Tractado de las drogas, y medicinas de las Indias Orientales, con sus plantas debuxadas al bivo por Cristóval Acosta medico y cirujano que las vio ocularmente. Martin de Victoria impressor de Su Magestad, Burgos. Biblioteca de la Universidad Complutense de Madrid.

Ahmed, M., Ting, I. P., & Scora, R. W. (1994). Leaf oil composition of *Salvia hispanica* L. from three geographical areas. *Journal of Essential Oil Research, 6*(3), 223–228.

Athar, M., & Nasir, S. M. (2005). Taxonomic perspective of plant species yielding vegetable oils used in cosmetics and skin care products. *African Journal of Biotechnology, 4*(1), 36–44.

Ayerza, R. (2011). Seed yield components, oil content, and fatty acid composition of two cultivars of moringa (*Moringa oleifera* Lam.) growing in the Arid Chaco of Argentina. *Industrial Crops and Products, 33*(2), 389–394.

Bazile, D. (2014). *Estado del arte de la quinua en el mundo en 2013*. Montpellier: FAO, Santiago de Chile and CIRAD.

Bazile, D., Pulvento, C., Verniau, A., Al-Nusairi, M. S., Ba, D., Breidy, J., Hassan, L., Mohammed, M. I., Mambetov, O., Otambekova, M., & Sepahvand, N. A. (2016). Worldwide evaluations of quinoa: Preliminary results from post international year of quinoa FAO projects in nine countries. *Frontiers in Plant Science, 7*, 850.

Bech-Larsen, T., & Grunert, K. G. (2003). The perceived healthiness of functional foods: A conjoint study of Danish, Finnish and American consumers' perception of functional foods. *Appetite, 40*(1), 9–14.

Brunsø, K., Fjord, T. A., & Grunert, K. G. (2002). Consumers' food choice and quality perception. *The Aarhus School of Business Publ., Aarhus, Denmark*. MAPP Working Paper No. 77.

Bugge, A. B. (2015). Why are alternative diets such as "low carb high fat" and "Super healthy family" so appealing to Norwegian food consumer. *Journal of Food Research, 4*(3), 89.

Caetano, B., de Moura, N., Almeida, A., Dias, M., Sivieri, K., Barbisan, L. (2016). Yacon (Smallanthus sonchifolius) as a Food Supplement: Health-Promoting Benefits of Fructooligosaccharides. *Nutrients, 8*(7), 436

Cahill, J. P. (2003). Ethnobotany of chia, *Salvia hispanica* L.(Lamiaceae). *Economic Botany, 57*(4), 604–618.

Campos, D., Betalleluz-Pallardel, I., Chirinos, R., Aguilar-Galvez, A., Noratto, G., & Pedreschi, R. (2012). Prebiotic effects of yacon (*Smallanthus sonchifolius* Poepp. & Endl), a source of fructooligosaccharides and phenolic compounds with antioxidant activity. *Food Chemistry, 135*(3), 1592–1599.

Cisneros-Zevallos, L. A., Nunez, R., Campos, D., Noratto, G., Chirinos, R., & Arvizu, C. (2002). Characterization and evaluation of fructooligosaccharides on yacon roots (Smallanthus sonchifolia Poepp. & End.) during storage. Abstr. Sess. 15 E. Nutraceuticals and functional foods. *Ann. Meet. Food Expo-Anaheim, California: 15E–27.*

Coates, W. & Ayerza, R. (1996). Production potential of chia in northwestern Argentina. *Industrial Crops and Products, 5*(3), 229–233.

Coelho, M. S., & Salas-Mellado, M. M. (2014). Chemical characterization of chia (*Salvia hispanica* L.) for use in food products. *Journal of Food and Nutrition Research, 2*(5), 263–269.

da Silva, M. A. S., de F. Hidalgo, A., de Morais, L. A. S., dos A. Gonçalves, M., & da Silva, S. M. P. (2002). Production of yacon plantlet (*Polymnia sonchifolia* Poep. et Endl.) in different organic fertilization. *International Conference on Medicinal and Aromatic Plants. Possibilities and Limitations of Medicinal and Aromatic Plant, 576*, 285–287.

de Almeida Paula, H. A., Abranches, M. V., & de Luces Fortes Ferreira, C. L. (2015). Yacon (*Smallanthus sonchifolius*): A food with multiple functions. *Critical Reviews in Food Science and Nutrition, 55*(1), 32–40.

Delatorre-Herrera, J. (2003). Current use of quinoa in Chile. *Food Reviews International, 19*(1–2), 155–165.

Delgado, G. T. C., Tamashiro, W. M., & Pastore, G. M. (2010). Immunomodulatory effects of fructans. *Food Research International, 43*(5), 1231–1236.

European Food Information Council (EUFIC) (2013). Annual report 2012. Resource document. https://www.eufic.org/images/uploads/files/EUFIC_Annual_Report_2012.pdf

Falguera, V., Aliguer, N., & Falguera, M. (2012). An integrated approach to current trends in food consumption: Moving toward functional and organic products? *Food Control, 26*(2), 274–281.

Fitschen, P. (2015). Are superfoods really "Super"? Resource document. http://www.bodybuilding.com/fun/are-superfoods-really-super. Accessed 18 Jan 2019.

Genta, S. B., Cabrera, W. M., Grau, A., & Sánchez, S. S. (2005). Subchronic 4-month oral toxicity study of dried *Smallanthus sonchifolius* (yacon) roots as a diet supplement in rats. *Food and Chemical Toxicology, 43*(11), 1657–1665.

Grau, A., & Rea, J. (1997). Yacon. *Smallanthus sonchifolius* (Poepp. And Endl.) H. Robinson. *Andean roots and tubers: Ahipa, arracacha, maca and yacón*, pp. 199–242.

Groeniger, J. O., van Lenthe, F. J., Beenackers, M. A., & Kamphuis, C. B. (2017). Does social distinction contribute to socioeconomic inequalities in diet: The case of 'superfoods' consumption. *International Journal of Behavioral Nutrition and Physical Activity, 14*(1), 40.

Hermann, M., Freire, I., & Pazos, C. (1999). Compositional diversity of the yacon storage root. CIP Program Report 1997e1998. Resource document. http://www.cipotato.org/roots-and-tubers/yacon. Accessed 19 May 2011.

Hondo, M., Okumura, Y., & Yamaki, T. (2000). A preparation method of yacon vinegar containing natural fructooligosaccharides. *Nippon Shokuhin Kagaku Kogaku Kaishi= Journal of the Japanese Society for Food Science and Technology, 47*(10), 803–807.

Jiménez, M. E., & Sammán, N. (2014). Chemical characterization and quantification of fructooligosaccharides, phenolic compounds and antiradical activity of Andean roots and tubers grown in Northwest of Argentina. *Archivos Latinoamericanos de Nutrición, 64*(2), 131–138.

Kaur, N., & Singh, D. P. (2017). Deciphering the consumer behaviour facets of functional foods: A literature review. *Appetite, 112*, 167–187.

Kleiman, R., Ashley, D. A., & Brown, J. H. (2008). Comparison of two seed oils used in cosmetics, moringa and marula. *Industrial Crops and Products, 28*(3), 361–364.

Lalas, S., & Tsaknis, J. (2002). Characterization of *Moringa oleifera* seed oil variety "Periyakulam 1". *Journal of Food Composition and Analysis, 15*(1), 65–77.

Leitão, R. F. F., Pizzinatto, A., Vitti, P., Shirose, I., & Mori, E. E. M. (1984). Estudos de duas cultivares de triticale e sua aplicação em produtos de massas alimentícias (macarrão, biscoito e bolos). *Boletim ITAL, 21*(3), 325–334.

Lutz, M. & Bascuñán-Godoy, L. (2017). The Revival of Quinoa: A Crop for Health. http://dx.doi.org/10.5772/65451

Maldonado, S., & Singh, J. D. C. (2008). Efecto de gelificantes en la formulación de dulce de yacón. *Food Science and Technology, 28*(2), 429–434.

Manrique I, Párraga A, & Hermann M. (2005). *Yacon syrup: Principles and processing. Series: Conservación y uso de la biodiversidad de raíces y tubérculos andinos: Una década de investigación para el desarrollo (1993–2003). No. 8B*. International Potato Center, Universidad Nacional Daniel Alcides Carrión, Erbacher Foundation, Swiss Agency for Development and Cooperation, Lima.

Martínez, E. A., Veas, E., Jorquera, C., San Martín, R., & Jara, P. (2009). Re-introduction of quinoa into Arid Chile: Cultivation of two lowland races under extremely low irrigation. *Journal of Agronomy and Crop Science, 195*(1), 1–10.

Mattson, F. H., & Grundy, S. M. (1985). Comparison of effects of dietary saturated, monounsaturated, and polyunsaturated fatty acids on plasma lipids and lipoproteins in man. *Journal of Lipid Research, 26*(2), 194–202.

Mintel. (2016). Resource document. https://www.mintel.com/press-centre/food-and-drink/super-growth-for-super-foods-new-product-development-shoots-up-202-globally-over-the-past-five-years. Accessed 15 Feb 2019.

Moscatto, J. A., Borsato, D., Bona, E., De Oliveira, A. S., & de Oliveira Hauly, M. C. (2006). The optimization of the formulation for a chocolate cake containing inulin and yacon meal. *International Journal of Food Science & Technology, 41*(2), 181–188.

NRC (National Research Council). (1989). *Lost crops of the Incas: Little-known plants of the Andes with promise for worldwide cultivation* (p. 415). Washington, DC: National Academy Press.

Ohyama, T., Ito, O., Yasuyoshi, S., Ikarashi, T., Minamisawa, K., Kubota, M., Tsukihashi, T., & Asami, T. (1990). Composition of storage carbohydrate in tubers of yacon (*Polymnia sonchifolia*). *Soil Science and Plant Nutrition, 36*(1), 167–171.

Ojansivu, I., Ferreira, C. L., & Salminen, S. (2011). Yacon, a new source of prebiotic oligosaccharides with a history of safe use. *Trends in Food Science & Technology, 22*(1), 40–46.

Pedreschi, R., Campos, D., Noratto, G., Chirinos, R., & Cisneros-Zevallos, L. (2003). Andean yacon root (*Smallanthus sonchifolius* Poepp. Endl) fructooligosaccharides as a potential novel source of prebiotics. *Journal of Agricultural and Food Chemistry, 51*(18), 5278–5284.

Pelchat, M. L., Johnson, A., Chan, R., Valdez, J., & Ragland, J. D. (2004). Images of desire: Food-craving activation during fMRI. *NeuroImage, 23*(4), 1486–1493.

Pineiro, M., Asp, N. G., Reid, G., Macfarlane, S., Morelli, L., Brunser, O., & Tuohy, K. (2008). FAO technical meeting on prebiotics. *Journal of Clinical Gastroenterology, 42*, S156–S159.

Prati, P., Berbari, S. A. G., Pacheco, M. T. B., Silva, M. G., & Nacazume, N. (2009). Estabilidade dos componentes funcionais de geleia de yacon, goiaba e acerola, sem adição de açúcares. *Brazilian Journal of Food Technology, 12*(4), 285–294.

Pulvento, C., Riccardi, M., Lavini, A., d'Andria, R., Iafelice, G., & Marconi, E. (2010). Field trial evaluation of two *chenopodium quinoa* genotypes grown under rain-fed conditions in a typical Mediterranean environment in South Italy. *Journal of Agronomy and Crop Science, 196*(6), 407–411.

Respondek, F., Gerard, P., Bossis, M., Boschat, L., Bruneau, A., Rabot, S., Wagner, A., & Martin, J. C. (2013). Short-chain fructo-oligosaccharides modulate intestinal microbiota and metabolic parameters of humanized gnotobiotic diet induced obesity mice. *PLoS One, 8*(8), e71026.

Roberfroid, M., Gibson, G. R., Hoyles, L., McCartney, A. L., Rastall, R., Rowland, I., et al. (2010). Prebiotic effects: Metabolic and health benefits. *British Journal of Nutrition, 104*(S2), S1–S63.

Rolim, P. M. (2015). Development of prebiotic food products and health benefits. *Food Science and Technology, 35*(1), 3–10.

Rosa, C. S., Oliveira, V. R., Viera, V. B., Gressler, C., & Viega, S. (2009). Elaboracao de bolo com farinha de Yacon. *Ciencia Rural, 39*, 1869–1872.

Ruiz, K. B., Biondi, S., Oses, R., Acuña-Rodríguez, I. S., Antognoni, F., Martinez-Mosqueira, E. A., Coulibaly, A., Canahua-Murillo, A., Pinto, M., Zurita-Silva, A., & Bazile, D. (2014). Quinoa biodiversity and sustainability for food security under climate change. A review. *Agronomy for Sustainable Development, 34*(2), 349–359.

Rützler, H., & Reiter, W. (2015). Food Report (2016). Resource document. https://onlineshop. zukunftsinstitut.de/shop/food-report-2016-en/. Accessed 18 Oct 2018.

Santana, I., & Cardoso, M. H. (2008). Raiz tuberosa de yacon (*Smallanthus sonchifolius*): potencialidade de cultivo, aspectos tecnológicos e nutricionais. *Ciência Rural, 38*(3), 898–906.

Tavares, L. S., Junqueira, L. A., de Oliveira Guimarães, Í. C., & de Resende, J. V. (2018). Cold extraction method of chia seed mucilage (*Salvia hispanica* L.): Effect on yield and rheological behaviour. *Journal of Food Science and Technology, 55*(2), 457–466.

Vazquez-Ovando, A., Rosado-Rubio, G., Chel-Guerrero, L., & Betancur-Ancona, D. (2009). Physicochemical properties of a fibrous fraction from chia (Salvia hispanica L.). *LWT-Food Science and Technology, 42*(1), 168–173.

Vega-Gálvez, A., Miranda, M., Vergara, J., Uribe, E., Puente, L., & Martínez, E. A. (2010). Nutrition facts and functional potential of quinoa (*Chenopodium quinoa* willd.), an ancient Andean grain: A review. *Journal of the Science of Food and Agriculture, 90*(15), 2541–2547.

Vilhena, S. M. C., Câmara, F. L. D. A., & Kakihara, S. T. (2000). O cultivo de yacon no Brasil. *Horticultura Brasileira, 18*, 5–8.

Vio, F., & Uauy, R. (2007). Food policy for developing countries: Case studies. In P. Pinstrup-Andersen, & F. Cheng (Eds.) No. 9–5; Resource document. http://go.nature.com/prjsk.

Voragen, A. G. (1998). Technological aspects of functional food-related carbohydrates. *Trends in Food Science & Technology, 9*(8–9), 328–335.

Wyatt, S. B., Winters, K. P., & Dubbert, P. M. (2006). Overweight and obesity: Prevalence, consequences, and causes of a growing public health problem. *The American Journal of the Medical Sciences, 331*(4), 166–174.

Yun, E. Y., Kim, H. S., Kim, Y. E., Kang, M. K., Ma, J. E., Lee, G. D., Cho, Y. J., Kim, H. C., Lee, J. D., Hwang, Y. S., & Jeong, Y. Y. (2010). A case of anaphylaxis after the ingestion of yacon. *Allergy, Asthma & Immunology Research, 2*(2), 149–152.

Zurita-Silva, A., Fuentes, F., Zamora, P., Jacobsen, S.-E., & Schwember, A. E. (2014). Breeding quinoa (*Chenopodium quinoa*): Potential and perspectives. *Molecular Breeding, 34*, 13–30.

Part II
Food Tech, Society and Industry

Chapter 6
Altering Production Patterns in the Food Industry: 3D Food Printing

Ioannis Skartsaris and Cinzia Piatti

6.1 Introduction

Throughout history, technological innovation has been one of the driving forces behind all major shifts in production and consumption patterns. Among these, digitalisation had met with a significant upturn during the last century and has resulted in unparalleled precision and efficiency in mass manufacturing processes. Consequently, this paved the way for an overvamp in the wealth accumulation models of the present. These technologies enabled global manufacturing to further specialize and offer products that contrast with fordist production patterns which were mostly focused on massive and unified manufacturing to feed the needs of the least possible segmented market (Hollingsworth and Boyer 1997). In the dawn of a post-fordism and a flexible specialization production shift of the global industrial structure, most manufacturing fields took advantage of the benefits specialization in production held in store. Diversification was the new norm and niche markets addressed the new profit- making drivers of the industry (Hollingsworth and Boyer 1997). This was valid not only for the manufacturing industry, but also in the food-related production, both at the field level and in the processing industry. Among many layers of enquiry in agrifood, one that is currently under-researched is meal creation. A big share of the research is occupied by research on production sites but less is devoted to the following steps along the entire food provisioning chain. What happens to food products after the farmgate and before getting on the shelves or before being consumed is a still long path to walk, and is subject to scrutiny too. Specifically, we have started wondering what sort of tech innovations that can have a disruptive character not only at the industrial level but also in social terms are currently available, and we have focused on 3D food printing as the new promising one,

I. Skartsaris · C. Piatti (✉)
Department of Societal Transition and Agriculture, University of Hohenheim,
Stuttgart, Germany
e-mail: cinzia.piatti@uni-hohenheim.de

© Springer Nature Switzerland AG 2019
C. Piatti et al. (eds.), *Food Tech Transitions*,
https://doi.org/10.1007/978-3-030-21059-5_6

already adopted at the industrial level and moved to the household level too, because it is paradigmatic of long-term issues around production. In Chap. 7 a dissertation on consumption politics proposes that the industrial level seems to be evolving and introducing market and production spaces which will thus create new niche spaces for food technologies to be employed at the household level. 3D food printing advocates market it as a panacea for a lot of problematic issues we are familiar with: for instance, they sustain that 3D food printing has the potential to reverse the severe impacts of traditional food production in the environmental and humanitarian field (Lin 2015). Solutions could include a wider implementation of entomophagy and supply chain optimization through the diminishment of the need for extended retailing, packaging and transportation (Mohr and Khan 2015). Again, advocates claim that it also holds the potential to implement ambitious processes like bioprinting (Godoi et al. 2016), the lab cultivation of animal tissue following in-vitro techniques to generate artificial animal tissue (Fountain 2013). Afterwards, the cultivated tissue can be three dimensionally (3D) printed in structures and shapes resembling common food like a burger (Fountain 2013). All these make up an endeavoring mix of possible applications. 3D food printing can be classified as the last major innovation that could be of the greatest potential to individual creationism (Mohr and Khan 2015) and has already been partly incorporated in contemporary confectionary manufacturing in the food industry, with sizable efforts also made by the pioneering agents to further expand the field of implementation to new directions. Does 3D food printing hold any true potential to be efficiently diffused and spread in the directions of environmental improvements or nutritional benefits? What sort of implications for industrial patterns of production? Here, in this chapter, we want to analyze the implications its implementation might have, specifically in the meal creation area in which additive technology and 3D food printing plays a curious role, as for many advocates 3D food printing has a disruptive potential. The reason why it is interesting for us is that, in political economy terms, this can be translated as reversing industrial patterns, specifically what part of the scholarship refers to as post-fordist ones (e.g. Lipietz 1997). Whereas in a manifold of industries 3D printing has enabled the diffusion of revolutionary technologies like "rapid prototyping" (T N O Crisp 2014), "build to order" (Sun et al. 2015) and the most efficient just-in time-mechanisms, in the area of food design and manufacturing things have remained rather traditional and subject to minimal change due to the complex nature of the medium. Production has remained manually driven and digitalization or major mechanical implementations negligible. 3D food printing constitutes a new norm that attempts to incorporate all these approaches in fields that have so far not yet being realized.

6.2   On Production Patterns and 3D Food Printing

Technology and innovation have been central for the food industry development since its inception. One potential segmentation could be the implementation in three major market levels which, according to Pinna et al. (2016), are mass production on the industrial scale, domestic production on the household level, business to business and food artistry level. A proper framework to introduce any potential disruption in food accumulation patterns which seems quite apt for understanding the change of production patterns is an analogy to transcend from fordism accumulation industry formations to post-fordism regimes. 'Fordism' refers to a framework of accumulation that features standardized mass production principles, drawing advantage from efficient labor allocation, utmost specialization and precise processes which result in immensely consumed products appealing to mass demand with very little individualism in adaptation (Burrows et al. 1992). In many aspects of the industry where technological inputs have proven crucial to overhauling production processes, following the decline of fordism production patterns new accumulation regimes were slowly starting to be acknowledged and sketched out by scholars and researchers. The main characteristics of post-fordism regimes are the shift of fixed industrial processes to more flexible forms of production and accumulation of capital (Ackroyd and Thompson 2006). Post-fordism therefore for some is a contemporary accumulation framework that functions as a vector of interactions between technological inputs, institutional regulation, and political governance, generally narrated by strategic choice (Hirst and Zeitlin 1991), with specific characteristic of high degree of mobility (of capital, commodities and labour) and high levels of specialization and competition. It can be regarded as a sought out 'competitive strategy' that draws advantage of competitive employees and state-of-the-art research to yield multitudes of applications and adapt rapidly to changes in demand and the general marketplace. The principal archetype of production is highly skilled labor that results in small batch production highly diversifiable according to individual taste and requirements (Smith 1989). The introduction of new technologies to bring such reformations into effect and realize deep market segmentation through highly customized designs and production is represented in the word 'flexible', whereas 'specialization' signals the end of fordism mass frameworks of production for adaptive and elastic ones (Smith, 1989). More importantly, an important feature is a predominant focus on consumption (Morozov 2019), which causes a shift of attention to this latter at the expenses of actors and materials employed in production processes whose issues remain mostly unsolved. Consumption practices are important to consider if we are to face transition to sustainability, because they address contemporary issues which range from wrong resources allocation to waste to alienation; but such a change in focus, while marking a problematic characteristic of our contemporary age in a classic Gramscian fashion on false consciousness issues, also leaves production (and power) relations unattended.

Mass production has evolved into a strange mix of the extreme fragmentation and market diversification to solve market saturation. Whereas technology has helped propel heavy industries to flexible production patterns, mass meal production had never met with technological innovations that could propel it substantially. Mass production of highly demanded meals like pizzas, crackers, and cookies is the new norm in research and feasible implementation of 3D food printing solutions, in this area most sought-after by innovative bodies. At the industrial end, feasible implementations have met with greater success than their counterparts at the household level. Presently, 3D food printing could be rather classified as an industrial application with economies of scale being the only viable path to feasibility. At the household level, there is a number of stakeholders spanning from start-ups to small companies that research and develop in various levels, the most promising of which is shrinking down 3D food printers to ordinary household appliances that can cover a wider spectrum of implementations. The diffusion of such an innovative technology in the all-important front of global nutrition, cannot be expected to arrive without ample changes of economic and societal nature. The high diversity of the actors involved during the infant steps the technology currently finds itself into further complicates the *collage* of conveyed information with motives behind each study and the expected results, diversifying substantially along the spectrum of researching bodies which may consist of anything between multinational corporations, innovative start-ups and the academia, all the way up to food activists and the various agents of food movements worldwide (Lupton and Turner 2017).

6.3 3D Food Printing

3D food printing involves many processes where ingredients such as powders and miscellaneous liquids are stacked up layer by layer through a variety of methods and combinations to achieve tridimensional edible solid structures (Lipton et al. 2016). 3D food printing can be classified as a sub-branch of a wider sum of 3D printing technologies that are widely referred to as 'additive manufacturing' or alternatively 'freeform fabrication' (a bit less frequent is the use of rapid manufacturing and rapid prototyping). The term 'additive manufacturing' refers to computer-aided designs (CAD) replication of solid 3D projects (Pinna et al. 2016), with the solid form end product bearing resemblance to most geometrical shapes and in the particular implementation of 3D food printing, firmly brought together through the process described above to be similar to all properties and traits of a traditional food product (Severini and Derossi 2016). The necessary designs and commands are being extracted from corresponding databases and used to reproduce original designs in identical copies. This, in essence, means that the patterns design and decorations are reproducible by anyone who enjoys access to the database and can be effortlessly shared across the globe. 3D printing, in general, has enjoyed integration in an abundance of technological applications in miscellaneous fields and industries. It operates in three modes around fusion, disposition and cutting, each specifically used for different materials so that basically the materials used will, respectively, being fused, or used for extrusion or for creating layers that will stick together.[1] 'Soft-materials' are employed for

---

[1] "Controlled fusion" is the mode where photocurable liquids have their surface layers fused by either heat or light sources in different rounds with unfused materials discarded after construction. "Controlled Disposition" mimics inkjet-type printing processes by utilizing either extrusion nozzles or inject-type print heads with the aim of extruding a constant flow of material to structure solid shapes by extracting layer upon layer. Different streams can also be mixed and fortify structure formations. Furthermore, coloring and binders can be also printed upon powder layers and deposited uninterruptedly. "Controlled cutting with Lamination" refers to the process of creating sheets of basic materials in specific shapes that result in layers stacked together with bonding materials. All additive manufacturing technologies can be combined in any way (Wegrzyn et al. 2012).

their extrusion characteristics. In this approach, edible constituents are stacked up to a 3D solid form. The most crucial attribute for the selection of materials is viscosity, which has to remain both low in order to be successfully extruded from the nozzle as well as above certain thresholds in order for the product to remain solid and firm (Godoi et al. 2016). Lastly, Selective Laser Sintering (SLS) is one of the predominant ways to achieve 3D manufacturing by employing CAD designs. The process involves intensive heat up of the powder by laser beams up to the melting point in order for them to get sintered together up to the final product assembly. The main argument in favor of SLS is the resource savings it offers. Up to now, it remains the highest among all 3D printing technologies (Wellington 2014).

In food production, additive manufacturing classifies food materials into three main categories according to the usability degree they present. The first group includes materials that are "natively printable" and consists of all physicochemical traits that enable ease of application in 3D food printing scenarios. This means they possess the necessary stability traits to hold form and structure after extrusion without the need for further processing afterwards. "Non-traditionally printable" materials, secondly, require additional post-cooking processes to come up with the same effect. "Non-printable" materials, on the other hand, can in no way be engaged without any hydro-colloid addition and meticulous processing (Pinna et al. 2016). Solid freeform fabrication (SFF) is the process of constructing solid three-dimensional objects conceived and designed by digital means without human labor at any part (Bourell et al. 1990). It is further classified into three methods: Stereolithography Lasing, Fused Deposition Modelling, and Selective Laser Sintering.[2] By employing either of these, 3D food printing can yield unparalleled precision processes that could have never been achieved through human labor. Precise cutting and drilling to form solid shapes and structures constitute the major achievements to date. Another common food printing application is the usage of stereolithography to turn egg whites into different patterns through laser technology. Additionally, solid sugar sculptures are producible by ink–jetting binders to achieve solid 3D-formations (Lipton et al. 2016). According to Sun et al. (2015) a key characteristic of 3D food printing is the possibility to employ innovatory materials like insects, algae or seaweeds that respond to these main characteristics and enhance meals. Apart from warm meals, 3D food printing has been met with success in producing frozen foods and foodstuff. By introducing a cooling approach based on liquid nitrogen, ice cream can be 3D printed and instantly consumed. The concept is called "foodForm" and, apart from ice cream, it can be employed on a manifold of food types, ranging from solid food like sugar products, as well as dough food and eggs (Yang et al. 2017). Three dimensional solid and complex chocolate forms by using liquid, soft or easily modifiable material (Severini and Derossi 2016) have already been incorporated in the industry through processes that are precise and accomplishable in the industrial level. All these can lead to impressive cost savings and processes with utmost efficacy.

---

[2] The entire spectrum of technologies encompasses so far inkjet 3D printing, powder bed printing, stereolithography, selective laser sintering, direct metal laser sintering, electron beam melting, fused deposition modeling and laminated object manufacturing (Wegrzyn et al. 2012).

6.4   Promises of 3D Food Printing

The identity of the major industrial stakeholders in the field presently is mainly comprised of startups and small companies that drive up the innovation and undertake the greater risks associated with the research and development. There is a handful of enterprises stretching in few places in the Global North (specifically Europe, US, Netherlands). When researching, we came up with few names currently active on the market. Their products interest mainly the industry, some the military complex, few the catering sector and the households. These players have different interests and targets, but they all long legitimation. The legitimation we refer to is to be found in the social issues currently taking space in our newspapers and public debates, and as such advocates of 3D food printing reverse to these same issues. For instance, Tran (2016) cites potential to reduce climate change by shifting agricultural production to accommodate for 3D food printing ingredients instead of miscellaneous food varieties and hence transform to a more sustainable and environmentally friendly practice. Lupton and Turner (2017) list food sustainability, food waste, ethical consumption, environmental degradation and world hunger issues, as fields of impact where 3D food printing is expected by food activists to positively engage and improve ratings and performance. Reducing costs, and consequentially the environmental burdens, is one principal attribute Nair (2016) considers for 3D food printing to carry, beyond addressing nutrient deficits by personalized diets and targeted ingredients. Going one step further, the author puts hope in tackling malnutrition efficiently through the use of additive technology and pave the way for healthier food. This should grab the interest of several private and public agents under the scope of improving food security. Sun et al. (2015) view 'high-value, low-volume customisation' as a realistic picture from the future that 3D food printing brings to the culinary sector with its potential to serve customers' needs adequately, whilst maintaining the present way of living. It is a route separating from the norm of today that salvages the hindrances of today where highly sophisticated designs and custom food yields increased costs of production for very low quantities, which remain limited in aesthetics and design options. On a more individual level, 3D food printers, its advocates say, will provide the user with a multitude of new resources to experiment and tailor production in revolutionary ways through increased ease of use and less complexity; this would also provide innovative designs and unparalleled customization in contemporary and future food (Sun et al. 2015). Personalized nutrition or "Consumer-Designed Food Fabrication" as also referred by Wegrzyn et al. (2012) is a major field in which 3D food printing promise to deliver customization and healthy derivatives of contemporary meals but also address the needs of segments of the population requiring particular food production approaches not yet realizable in the food industry. Finally, for Wegrzyn et al. (2012) 3D food printing would enable consumers to take food design and production in their own hands and influence and guide it according to their personal taste or nutritional needs. Every step of the process can be tailorable and customizable accordingly, whilst opening every possibility for online collaboration.

6.5 Printing Controversies

Much of the applications of 3D food printing happen in the confectionery industry and those industries which use carbohydrates and sugar-based ingredients, because of the characteristics of the ingredients themselves, highly desirable for industrial application and specifically for 3D food printing (as seen in the first section, such as physicochemical viscosity, extrusion, fusion, absorbency and thermo-electrical conductivity). But there are two specific industries in which 3D food printing is employed which might create tensions. In fact, in a very futuristic fashion, 3D food printers could bring about the integration of alternative ingredients to enhance new food and bring forth new meals and flavors alongside environmental salvage, but have some issues to overcome, two of which are the most interesting for us. These are also interestingly going to have an impact on the issues mentioned above, and very likely affect the food choices of individuals. The first is amongst the most controversial aspects to combine with 3D food printing, and that is entomophagy; this term refers to a diet consisting of insect consumption which benefits nutritional as well as environmental scope.[3] Insects have long been a key part in the traditional diets of many eastern civilizations. Some species are even consumed in the western countries like the US, Spain, France, and Italy and in total, 1900 species are part of the everyday diet of about 2 billion people (FAO 2013). Ingredients or whole food comprised entirely out of insects can be incorporated into western diets through additive technology in a personalized fashion compatible with individuality in nutritional needs and aesthetics (Luimstra 2014). Pinna et al. (2016) claims that 3D food printing can result in the introduction of healthier and nutritionally dense food compared to traditional food preparatory processes and ingredients. Insects can also constitute ingredients instead of purely raw material. They can be combined and mixed along with several ingredients like beet leaf, lupine seeds and grass for enhanced nutritional value and aesthetics (TNO Crisp 2014). For instance, 'Insects Au Gratin' is a project where insects as raw materials get combined with cheese

---

[3] Food processing innovations like food printing by using natural components of ecological value like these, can improve profit margins per hectare whilst supplement the absorbency and sustainability of future food (Sun et al. 2015). Insects hold a significant role in supporting waste biodegradation and plant reproduction. Insects carry traits of sustainable and easy cultivation while being abundant in protein and fat contents. The amounts of unsaturated omega 3 and fatty acids further surpass those found in fish and animals (FAO 2013). The cultivation deterrents are significantly far less than those of other practices and insect conversion rates can reach heights of weight-to- feed ratios 1–2 with cultivation carrying potential to flourish in both organic and conventional substrates. Also, entomophagy can lead to reduced ammonia and greenhouse emissions whilst remaining fruitful in cases of land constraints. Research has also shown an incompatibility of human hosts with zoonotic infections (FAO 2013). Consumption can, therefore, be potentially easier to implement in parts of the world where control mechanisms, hygiene, and the knowledge behind remain still constricted.

varieties, icing, and other compounds to offer a diversity of desirable flavors. Experimentation does not only involve material combinations but various technological approaches like multiple heads extrusion to radiate results (Sun et al. 2015). Insects can be also used as protein powder to further enhance nutritional quality and the protein content of the meal. Wheat based snacks and meals can benefit from supplementing on their poor protein content along with biological traits by implementing of 3D food printing solutions (Severini and Derossi 2016).

The second controversy is bioprinting. Bioprinting refers to the cultivation of stem cells outside the organism from which they have originally been extracted, transferred and reproduced in a suitable medium consisting of all sufficient substrates nutritional elements and conditions necessary for reproduction (Murphy and Atala 2014). It is referred to as the 'in vitro' technique. More specifically, once the stem cells have been transferred to the bioreactor, they undergo differentiation and start to mature in the form of muscle cells which are identical to naturally grown animal cells and totally edible. 3D printing implementation is the last step of the process (Murphy and Atala 2014). In order to form 3D structures, the cells are fabricated into layers which are then extruded on top of each other and bonded together by biochemicals and biological materials. There is even more than one approach to bio- printing. Self-assembly and biomimicry are two of the most prominent. Printing living tissue, however, in contrast to fabricating 3D structures with plastic - or even metal - materials, has proven to be considerably demanding. Living tissue is extremely sensible and fragile. Fabricating the extra-cellular matrix requires specific and delicate processes in order to adequately form the microarchitecture of the extracellular matrix (Murphy and Atala, 2014). Bioprinting materials, as is the case with all 3D printing ingredients, must adhere to specific requirements to stand eligible for 3D food printing success. They must be of specific viscosity and adhere to the extrusion standards necessary that extrusion nozzles require. Bioprinting of perfectly functioning human organs from scratch, nevertheless, is a breakthrough that has already enjoyed significant success (Lipson and Kurman 2013). Approaches of similar nature like 'biomimicry' or 'autonomous self-assembly' are the ones that have yielded the most significant results in the medical field (Murphy and Atala 2014). This paves hope for edible technologies to take off in the near future. As handling of tender biological materials evolves and a prosperous and necessary field as is one of the medical applications continuously requires further breakthroughs and improvements, bioprinting of edible products might develop alongside as the two practices share similar attributes and challenges. In a prototype developed in the Netherlands, as many as 20.000 strips of cultivated edible fibers needed to be utilized for a single burger to be artificially (re)produced.[4]

---

[4] Meat from bovines or livestock are not the only one kind of tissues which interest the industry; apparently Singapore-based company Shiok Meats will be hosting a tasting of dumplings made with its cell-based shrimp in March 2019 at the Disruption in Food and Sustainability Summit (https://arf.org.sg/dfss/) in their home country.

6.6   3D Food Printing Limitations

Although promising, food printing is most likely to remain unrealized in a similar degree during the near future (Nair 2016). For example, food engineering of complex food consisting of a multitude of ingredients like a burger is yet to be realized (Lipson and Kurman, 2013). The main factor hindering further adaptation is the specific requirements in specific parameters that all ingredients must adhere to in order to be utilized in a 3D food printer successfully. It is what makes complex food printing inapplicable on a multispectral level of ingredients and holds additive technology back from appealing to mass demand by manufacturing complex meals like a burger (Godoi et al. 2016). There is a manifold of physicochemical viscosity, absorbency and thermo-electrical conductivity ranges that have to be strictly respected in order for 3D food printing to function on a wider level. These properties show very high variation from batch to batch and this raises unpredictability significantly. Limited testing solely relying on pre-existing databases or theoretical modelling complicates the efficiency of the procedure even more (Sun et al. 2015). The University of Singapore (ibid.) has attempted to resolve this impasse by researching on a food printing platform with commercial aspirations that could associate the miscellaneous properties of the several ingredients and optimize fabrication processes. The final spectrum of the ingredients that can be utilized, however, remains severely limited. Many researchers doubt that implementing food printing in single step processes shall constitute additive technology's main aim. In their view, the focus should rather be the fabrication of revolutionary textures and supplementary nutritional value (Sun et al. 2015). On the basis of current research, many agrees that no matter how fascinating 3D printing up to the point of complete replication may be, it is quite unlikely that it will end up being a superior form of producing complex meals which are already out there in the short run. Even if such a feat is eventually achieved, it is still improbable it will prove possible on the industrial level when comparing to traditional manufacturing processes in the short term. Slightly outside the framework discussed here above, there are two considerations about the specific applications of entomophagy and bioprinting which make the picture even more blurred. At its present state, bioprinting faces severe limitations. The main hindrances that need be overcome are the steep investment costs, the prolonged maturation time-spans that may reach up to 7 days for a single product (Marga et al. 2012) and also "the spatial resolution of the final construct" (Murphy and Atala 2014). If these constraints could suddenly be overcome (and we stress 'if') advocates insist that significant amendments to meat production and distribution patterns worldwide would be proven. Such alterations would impact the entire meat supply chain worldwide by remapping supply chains in favor of developed urban centers against rural communities. Of course, in terms of land scarcity and environmental impact, decreased meat production would signal a great achievement. Facing severe land scarcity hindrances, farming in Singapore, as the authors mentioned above maintain, is a minor practice at the same time the population upsurges. But this mix brings forth severe sovereignty concerns with input dependencies rendering the survival of such urban ecosystems unfeasible. Secondly, and here in common with entomophagy, there are some social stigmas yet to be overcome. Advocates claim that both would benefit the masses with cultural, ethical, health or religious restrictions on meat consumption, as well as populations with restricted access to safe food consumption in the future. Alternative ingredients like insects, algae, duckweed, seeds or even in-vitro meat in the eyes of many researchers are expected to play major role in combatting world hunger (Lupton and Turner 2017) and benefitting environmental conservation through greater feed conversion, reduced greenhouse gases emissions, minimized risk of spread of diseases associated with animal husbandry, increased animal welfare and an optimal feed to food conversion ratio (FAO 2013).

6.7   Disrupting Production Patterns?

Enthusiastic claims are the essence of launching new products and technologies but from our perspective whether these new products, and the new technologies behind, will be successful is more than a matter of novelties, or of just simplifying some parts of everyday activities at both industrial and household level. 3D food printing holds promise to innovate accumulation patterns by disentangling developing countries from indulging in raw material exporting without having access to the means of converting such wealth to end products and acquire the high returns associated with an added value. From a capitalist perspective, issues to be overtaken can be disentangled quite easily from fordist to post-fordist patterns. To clarify it, we take the example of one specific commodity that comes from developing countries and is highly employed in the current development of 3D food printing: cocoa, highly used in the confectionery and pastry industry. Confectionary manufacturing at the end product of cacao production is labor-intense and adheres to fordism production patterns. Fordist industries are characterized by substantial entry barriers as economies of scale account for the greatest degree of financial sustainability. In developing countries, such entry barriers are hard to overcome - even harder in sustainable ways. Also, for a sustainable economic development, five types of sustainable capital are necessary – human, social, manufacturing and financial capital that all together incorporate into the wider natural capital of a country (Forum of the future 2018). In the case of developing countries which also enjoy the proper climatic attributes to undergo cocoa production, the natural capital adheres to all specific attributes that establish proper cacao production. The resources are there, and the climate properly regulated to make cacao production flourish. The social and human capital is also mainly comprised of agents orientated at cacao production with an adequate amount of knowledge and expertise. Though the human capital can be easy to come across (even more in this age of migration and displacement, where people with a job in their country of origin will hold a job desperately), and the natural capital abundant, the manufactured and financial capital in those cases is severely limited and constitutes a great deal of on-site production unfeasible for sustainable and prosperous high-end manufacturing of confectionery products. Going beyond traditional fordism paradigms to flexible specialization patterns could transcend the economic models of cocoa exporting countries to contemporary paradigms of small – batch production taking advantage of technological inputs and possibly constituting a big deal of the human capital involved, to white-collar workers. Fordism industries, being capital intensive are hardly applicable in many developing countries, but post-fordist ones have instead proved to be able to tackle these territorial boundaries. Even in less technology-intensive fields like food production, lack of access to the necessary financial capital is not the sole reason as one may guess. Intense investments are also stuck because of inefficient in many cases regulative actions, educational and justice institutions or governance mechanics that could support and regulate investments on a grand scale (Braun and Birner 2017). But from a political economy perspective, inherent issues such as those related to neo-colonialism or labour are not solved, as is the case from an ecological perspective, since plants before being a commodity are the expression of specific ecosystems and cannot be easily adapted to new needs or fashion, not to mention that ecological disruption can have unexpected consequences. 3D food printers realizable on the way the main innovators and scholars describe are expected to, bring forth mechanical efficiency and ease mass production through specialized and skilled labour consisting primary of engineering and design traits instead of traditional hand labor. This not only confirms the tendency of post-fordist systems to insist for a technical education with high engineering skills to be developed, it also confirms the trend of increased automation at the expenses of unskilled labour. Of course, the features highlighted for the example of the commodity cocoa can be held true for many other commodities, especially when coming from developing countries and being central for some countries' economies.

Getting into the specificity of production patterns, 3D food printing as an industrial solution can be deemed as one of the very first technologies of its kind that could possibly put a rather 'robotic' mass production pattern in track. There is a certain considerable degree that 3D food printing might be able to disrupt supply chains. The main expectations as encountered throughout most publications are the overhauling of food supply chains (Sun et al. 2015) which is something presented through the greater localization of food production by introducing build-to-order innovations. This is expected to be further assisted through greater networking and connectivity that will make the most of 'smart technologies' (Pinna et al. 2016) which can introduce an artificial intelligence factor in food production and logistics calculations to take advantage of modern computer processing power, to reach greater efficiency. Lupton and Turner (2017) support the argument of more efficient storage capacity and reduction in food waste and packaging to indirectly indicate how 3D food printing can help transform and alleviate several modern food supply chain inefficiencies. The main levels of massive food manufacturing like mass-produced food processed such as frozen pizzas or confectionaries, are labor intensive and consequently scale up through capital investments on the grand scale. Mohr and Khan (2015) consider that 3D printing is expected to totally overhaul resource efficiency by diminishing the complexity that production and manufacture frameworks carry. Resource efficiency in 3D food printing can be expected to arrive in the form of food waste reduction. 3D food printing is expected to radicalize modern food manufacturing following short on 3D printing manufacturing overhauls. More specifically, Mohr and Khan (2015) list 'complexity reduction' and 'decentralisation of manufacturing' as two impact areas for 3D printing approaches to disrupt the future supply chains. It is clear that 3D food printing is expected to de-align mass food manufacturing from the need to scale up through capital and labour-intensive investments. The upcoming mobility of food manufacturing is fairly safely expected to drive up uptake in less developed countries, wherein the present, raw material production is the main part of the supply chain they are mostly engaged at (see Chaps. 4 and 5 of this book). Successful projects to bring food printing implementation in difficult to reach areas and remote places, even in the form of interstellar travel, will imply the decentralization of food production. 3D printing is presented by Mohr and Khan (2015) as a way to rapid manufacturing in responses to national disasters in hard to reach places but whether and how this will happen, and more importantly how quick will the political response be, is a highly doubtful and debatable aspect. We can expect 3D food printing to follow short on the total sum of 3D printing technologies and offer similar traits and attributes. By the same token, 'mitigating the risks of obsolescence' as described by the same authors to be the outcome of breaching the gap between production and home markets, would be addressed by increased mobility in production, effective resource management, diminishing special needs and inherited cost efficiencies. All in all, these are all more easily said than done. Added production volumes on-site, stack up with diminished transportation costs and production approaches focused on raw materials instead of end products (Mohr and Khan 2015). 3D food printing would enrich traditional food production with (ideally) skipping the need to undergo extensive post-processing steps such as cooking or baking (Godoi et al. 2016). This is inevitably expected to lead to significant cost reductions, better time efficiency and diminished need for manual labour, but the infant steps in which this industry is currently in cannot provide sustained reason to make it the case. Similar technological initiatives have narrated the transition to post-fordism transformations in heavy industries in the past sketching out obvious potential for 3D food printing to impact the food industry accordingly. The strong interest of the military complex in stakeholder involvements are to be expected to drive up the uptake of the technology through ambitious initiatives which though not signalling a greater industrial upshift, clearly constitute wider exposure for the technology.

Supplementary to mass production, the second pillar of transformation 3D food printing can bring about to food manufacturing - and bears the closest resemblance to post-fordism production models shift in the past - can be expected to be flexible nutritional qualities that are ever more achievable through the technology (Godoi et al. 2016) to which applicability is another interesting attribute of those technologies where targeted nutritional qualities are becoming ever easier to introduce through contemporary designs and exotic materials. By enabling greater ease of targeting diversifiable audiences in greater efficacy, 3D food printing promises to be diffused as a technological innovation in higher velocities by harvesting added value in profitable markets and being governed by swift changes in demand. Despite this, a very significant factor to accumulating cost savings through 3D food printing that remains underexposed at the moment is the ability to innovate textures and mouthfeels without necessarily mimicking the exact food itself, but rather building similar qualities through tailored mouthfeels. Cohen et al. (2009) have managed to achieve a broad range of mouthfeels only through usage of xanthan and gelatin as the main ingredients and have ultimately concluded, there is big potential in skipping the need to reconstruct traditional food for researching and 3D printing simulations of each particular meal instead. The goal is that replicating the mouthfeel through basic ingredients as building blocks a blind test may make the original and the simulated model indistinguishable. In addition, the fact that some ingredients of 3D food printing currently include insect parts such as insect flours does not represent yet a reason to be optimistic, because often customers ignore the composition of their meals. These remain the biggest obstacles, currently unresolved as eradicated in the very nature of relations of productions, which cannot testimony positively for a much needed change in production patterns.

References

Ackroyd, S., & Thompson, P. (2006). *All quiet on the workplace front? A critique of recent trends in British industrial sociology.* In: Patterns of Work in the Post-Fordist Era: Fordism and Post-Fordism. Cheltenham: Edward Elgar.

Bourell, D. L., Beaman, J. J., Marcus, H. L., & Barlow, J. W. (1990). Solid freeform fabrication an advanced manufacturing approach. *In Proceedings of the SFF Symposium (pp. 1–7).*

Braun, J., & Birner, R. (2017). Designing global governance for agricultural development and food and nutrition security. *Review of Development Economics, 21*(2), 265–284.

Burrows, R., Gilbert, N., & Pollert, A. (Eds.). (1992). *Fordism and flexibility: Divisions and Chang.E.* New York: St. Martin's Press.

Cohen, D. L., Lipton, J. I., Cutler, M., Coulter, D., Vesco, A., & Lipson, H. (2009). Hydrocolloid printing: A novel platform for customized food production. In H. L. Marcus (Ed.), *Proceedings of the solid freeform fabrication symposium* (pp. 807–818). Austin, TX: University of Texas.

FAO. (2013). Edible insects: future prospects for food and feed security (No. 171). Rome: Food and Agriculture Organization of the United Nations. (Edited by Van Huis, A., Van Itterbeeck, J., Klunder, H., Mertens, E., Halloran, A., Muir, G., & Vantomme, P.).

Forum of the future. (2018). The five types of capital. Resource document. https://www.forum-forthefuture.org/sites/default/files/project/downloads/five - capitals-model.pdf. Accessed 11 February 2018.

Fountain, H. (2013) At the printer, living tissue. *The New York Times.* Aug. 18, 2013. Resource document. http://www.nytimes.com/2013/08/20/science/next-out-of-the-printer-living-tissue. html?. Accessed 16 April 2018.

Godoi, F. C., Prakash, S., & Bhandari, B. R. (2016). 3d printing technologies applied for food design: Status and prospects. *Journal of Food Engineering, 179*, 44–54.

Hirst, P., & Zeitlin, J. (1991). State, democracy, socialism: Introduction. *International Journal of Human Resource Management, 20*(2), 133–137.

Hollingsworth, J. R., & Boyer, R. (1997). *Contemporary capitalism: The embeddedness of institutions.* Cambridge, UK: Cambridge University Press.

Lin, C. (2015). 3D food printing: A taste of the future. *Journal of Food Science Education, 14*(3), 86–87. https://doi.org/10.1111/1541-4329.12061.

Lipietz, A. (1997). The post Fordist world: Labor relations, international hierarchy and global ecology. *Review of International Political Economy, 4*, 1–41.

Lipson, H., & Kurman, M. (2013). *Fabricated: The new world of 3D printing*. Indianapolis, IN: Wiley.

Lipton, J. I., Cutler, M., Nigl, F., Cohen, D., Lipson, H. (2016). Additive manufacturing for the food industry. *Trends in Food Science and technology, 43*(1), 114–123.

Luimstra, J. (2014). The Future: A 3D Printed "Insects" Meal. Resource document. https://3dprinting.com/materials/edibles-food/future-3d-printed-insects-meal/.

Lupton, D., & Turner, B. (2017). 'Both fascinating and disturbing': consumer responses to 3D food printing and implications for food activism. In *Digital food activism* (pp. 151–167). London: Routledge. https://doi.org/10.4324/9781315109930_8

Marga, F., Jakab, K., Khatiwala, C., Shepherd, B., Dorfman, S., Hubbard, B., Stephen, C., & Forgacs, G. (2012). Toward engineering functional organ modules by additive manufacturing. *Biofabrication, 4*(2), 022001.

May, B. (2013). Bio-3d Printing *SIMTech Annual Manufacturing Forum 2013 Manufacturing Technologies: the next 20 years and beyond.*

Mohr, S., & Khan, O. (2015). 3D printing and its disruptive impacts on supply chains of the future. *Technology Innovation and Management Review, 5*(11), 20–24.

Morozov, E. (2019). Capitalism's new clothes. Resource document. https://thebaffler.com/latest/capitalisms-new-clothes-morozov. Accessed 5 February 2019.

Murphy, S., & Atala, A. (2014). 3D bioprinting of tissues and organs. *Nature Biotechnology, 32*(8), 773–785. Resource document. http://www.nature.com/doifinder/10.1038/nbt.2958%5Cn. http://www.ncbi.nlm.nih.gov/pubmed/25093879. Accessed 18 Nov 2018.

Nair, T. (2016). 3-D Printing for Food Security: Providing The Future Nutritious Meal. *RSIS Commentary 273.*

Pinna, C., Ramundo, L., Sisca, F. G., Angioletti, C. M., Taisch, M., & Terzi, S. (2016). Additive Manufacturing applications within Food industry: an actual overview and future opportunities. *In 21st Summer School Francesco Turco 2016* 18–24.

Severini, C., & Derossi, A. (2016). *Could the 3D printing technology be a useful strategy* to obtain customized nutrition? *Journal of Clinical Gastroenterology, 50*(December), S175–S178. Resource document. https://www.ncbi.nlm.nih.gov/pubmed/27741169.

Smith, C. (1989). Flexible specialization, automation and mass production. *Work, Employment and Society, 3*(2), 203–220.

Sun, J., Zhou, W., Huang, D., Fuh, J. Y., & Hong, G. S. (2015). An overview of 3D printing Technologies for Food Fabrication. *Food and Bioprocess Technology, 8*(8), 1605–1615.

TNO Crisp. (2014.) 3D Food printing Conference, *T N O Crisp. (2014). "3D Food Printing."*

Tran, J. (2016). 3D-printed food. *Minnesota Journal of Law Science & Technology J.L. Sci. & Tech, 17*, 855.

Wegrzyn, T., Golding, M., & Archer, R. H. (2012). Food layered manufacture: A new process for constructing solid foods. *Trends in Food Science and Technology, 27*(2), 66–72.

Wellington, M. (2014). How does selective laser sintering work? Resource document. http://3dprinthq.com/how-does-selective-laser-sintering- work/. Accessed 10 February 2018.

Yang, F., Zhang, M., & Bhandari, B. (2017). Recent development in 3D food printing. *Critical Reviews in Food Science and Nutrition, 57*(14), 3145–3153.

Chapter 7
Food Consumption and Technologies

Cinzia Piatti and Forough Khajehei

7.1 Introduction

Long time ago, the notion of a Great Transformation (Lee and Newby 1983: 26–39) made the nineteenth century Europe the centre of change in capitalist environment. The idea has to do with market economies and nation-states in which industrialization plays a central role. As Corrigan (1997:2) reminds us, though, those were the centuries of production, as the classic Marxist tradition imposes that the pivot around which all revolves (to be read as the economic, social and academic worlds) is production. So probably a new great transformation will entail consumption even more than what some thinkers might say. As reminded in the introduction chapter, consumption is now pivotal (Goodman and Dupuis 2002) and, consequently, affecting it can cause change in societal and economic organization more than a production patterns shifting. Arguably the main change in consumption happened in the 1950s of the last century; we know that the post-WWII imperatives of reconstruction were to produce more to avoid the communist perils (Patel 2013: 5) but this, as Campbell (1983) maintains, is counteracted by the fact that the Industrial Revolution involved a revolution in both production and consumption. In both of them, the role of technologies is paramount. Digitalisation, robotization, full automation are the imperative words of our time; robotics, genetics, blockchain technology, 3D printing and artificial intelligence have become familiar words, although the big audience might still not grasp all of them, and they are promised to change deeply our habits in the next decades. In writing this chapter we observed the many revolutions we see around us in everyday contexts about food technologies and food consumption.

C. Piatti (✉)
Department of Societal Transition and Agriculture, University of Hohenheim,
Stuttgart, Germany
e-mail: cinzia.piatti@uni-hohenheim.de

F. Khajehei
Institute of Crop Science, University of Hohenheim, Stuttgart, Germany

© Springer Nature Switzerland AG 2019
C. Piatti et al. (eds.), *Food Tech Transitions*,
https://doi.org/10.1007/978-3-030-21059-5_7

Whether we eat our own, self-produced food, cooked by ourselves or others in the same household, or we eat out more or less regularly, we are deeply immersed in the continued evolutions of food technologies, especially at the household level or in some public spaces such as catering places. We acknowledge that these simple acts are not only the result of our individual choices but depend on other factors. The pervasive character of technological transformations, of technology itself, makes it a quite exciting historical moment as well as quite worrisome, as they might change in unexpected ways the way we eat and consume, and since these activities do not happen in a vacuum, as Carolan (2012:281) reminds us, we can expect them to affect our personal, civic or professional relationships as well. The main critique to consumption is that it is a defining character of our age; currently though the environmental impact of our consumption styles has become an imperative, the quest for sustainability for which technology is considered to be the silver bullet, result in conflictual outcomes, as consumption has become an end in itself and is still is under scrutiny. Whether it is too early to talk about an effective disruption in our habits and ways of life, or whether this is already happening in a subtle way is beyond the scope of this chapter; in this chapter we want to reflect on food consumption and trends and scrutinize how some of the most promising new technologies and related trends might have an impact on them.

7.2   Consumption

Thompson (2016) argues that consumption of goods and services is embedded in our daily lives and we do not question it any longer, as it is considered as a given. If we take this as the consumer culture age, as Sassatelli proposes (2007) then we have to accept it as a defining feature of our modern lives; consumption highlights the range of social relations and interactions in which objects with specific functions are selected and used (Zelizer 2005); estimations of needs, wants and satisfaction are concomitant for addressing the role of consumption in modern industrial worlds and the role of status and reputation (Warde 2015:119). Food consumption, quite new in the realm of consumption theories, is analyzed from theoretical perspectives along the axis of culture, function, structure or development (Mennell et al. 1992; Holm 2013) and from empirical ones along culinary trends, class, ethnicity, religion, gender (Thompson, 1996; Germov and Williams 2010). Research has shown that there is a social differentiation in the way we consume food: for instance, Diner (2001) and Lupton (1996) have shown how social identities are defined by what we eat; Harrington et al. (2011) have shown how income and education affect food consumption, so for instance to abide to a healthy lifestyle and follow nutritional recommendations would be the result of a higher socioeconomic status. Little, though, has been said on the role of food technologies in the constitution, structuration or change of food consumption. Two main aspects are relevant for our understanding of food consumption and related role of technologies: one is the shift from home consumption to eating out, a recurrent and shifting theme in many ages; the second is the individualization that consumption seems to have undergone in the past decades.

1. Holm (2013) traces the change of food consumption from the industrial revolution, moving from numerous daily meals to the quite stabile number of three, as a consequence of having to adapt to new working conditions and the restructuring society underwent consequently. The twentieth century was the period of rapidly transforming the experiment-based knowledge and science of food processing and technologies into the contemporary industrial forms and rapidly reshaping the traditional methods into refined techniques, e.g. cooking was transformed into canning technology, sun drying methods were improved into more hygienic mechanized processing techniques such as convective hot air drying, cold storages were evolved into refrigeration and freezing (Truninger 2013). As we have detailed in another chapter in this book (Chap. 2) and taking the industrial revolution as watershed moment, the basic cooking tools were transformed and the rise of a new generation of food processing technologies may be noted as the moment that scientific achievements in fields of biology, chemistry or physics coupled with the technological advances offered by industrial revolution, and progressively prospered to the current food production and consumption chain. In this trend, food technologies contributed to relieve (mainly) women from household duties, among which food preparation was significant. The development and improvement in the field of food processing from the twentieth century till the present moment evolved continuously. The number of appliances introduced in households have increased to meet the different needs of changing habits. As a result of commodification, though, meals are consumed far from households, making the time for food preparation declining (Warde et al. 2007). Eating out, though, is constantly on the rise (Warde and Martens 2000). And still, according to Lin (2015), food preparation is expected to be over vamped as the procedures of transforming a certain set of ingredients to different textures and meals in virtually limitless possibilities, will be a driving force to key adoption. How this will unfold is the subject of our investigation, as there is an overlapping, or better a contamination, of practices and tendencies in each which has influenced the other and viceversa.

2. Warde (2015: 122) maintains that food consumption is a domain rapidly changing, exemplifying "the intertwining of the forces of globalization, commodification and aestheticization". In particular, he indicates the role of the 'foodie' as paradigmatic because it symbolizes the role of enthusiasm in consumer culture, the ability to elevate an ordinary activity into a core of the luxury industry and of distinction. In this enthusiasm, mass media and social media have played a central role in contributing to wide diffusion of images, lifestyle, a travel-culture. Veblen's theories about emulation (1994) are of course relevant, as consumption patterns are a territory in which human emulation is evident, although constrained by income, taste, education, culture, as the literature on consumption

has highlighted consistently (for a review see Paterson 2006). Grignon (1996), though, argues that post-industrial societies have undergone a deregulation following the post-industrial politics, transforming eating into an individualized and flexible activity, although research in different countries have shown this is not evident everywhere (Holm 2013:330). Particularly, Soron (2010:177) considers food consumption as part of a vast process of individualization and identity; more importantly in his argument it is the ethos of greening and sustainability, of which organic food is paramount, that helps explaining the evident turns in consumption as factor of identity. Soron is skeptical, though, of greening consumption; his point is that this radicalizes even more hedonistic consumption and paves the way to contradictory behaviour as it might induce more consumption. Schor's (2007) research is in line with this last consideration, as he expressed concern about the high material and environmental impact of consumption patterns. On the basis of this framework, in the next section we will discuss the last trends of food consumption according to the technological innovations currently available.

7.3   Trends

How do we make sense of the innovations that involve food consumption in relation to the technologies currently available? The issue of food tech innovation involves raw material processing, packaging, additives, forms of hyper- nutrition and taste through different consumption. To make sense of the kind of change technology is pinpointing we have looked at those which have received more attention by experts in the field or which look most promising from a sociological perspective. Out of the theoretical framework that we have delineated in the previous section through literature research, we have triangulated the tendencies in food consumption (commodification, globalization, aestheticization, individualization upon which we inscribe the health-related food consumption, sustainable consumption). We began with a desk research, comparing results obtained from online search engines about food consumption and food technologies with those obtained from literature research on same themes. That allowed us to align the important trends emerging in both scientific and lay/non-scientific fields; we focused on English-language entries in both fields and narrowed down to westernized countries. Initially we separated and differentiated between refrigeration systems and heating ones, and innovations for either household or the industry, with a further differentiation with food processing and the professional categories such as medium-to-large scale possible final users like bakeries, confectionery, catering industry or restaurants catering, or the food processing industrial sector. It has to be clarified that we did not categorize 'catering' and 'restaurants' in the main innovations found here following, because there has been sometimes an overlap of innovations moving from the catering and hospitality industry into households. Also, we have registered a distinction between mass-attended dining places, such as diners or fast-food, and the so-called 'fine-dining', since the products offered by the industrial actors here proposed provide for two different needs: one is mass-production, popular and cheap meals; and the other is artisanal, highly-skilled production, elitist and expensive meals, a highly contradicting category. Of course, the different targeted markets translate into a different manufactured end product, which addresses and responds to different sectors.[1] We noticed that there is no univocal direction in how different technologies are diffused and adopted, one sector influences the other in terms of trends, practices, demands, so there is no a hierarchical or unilinear stream for diffusion and/or adoption; the same overlapping happens for the functional categories such as refrigeration or storage, therefore we dropped the initial coding and created categories in which one or the other appear, so that it could be reflected the blurring of categories and orders previously clearly defined. Consequently, we have not made any specific separation between the innovations destined to household and those to professionals; much of the research of our colleagues in crop science will serve both the food processing industry, which specializes in products for both catering and final customers, and the supply chain such as global distribution, but although obvious differences in terms of scale and investments remain much of the changes in technological adoption and food consumption are blurred in different domains. For instance, the boundaries can become saturated between robotization, 3D food printing, full automation and what expands the equivalent stakeholder field of view further.

7.3.1 Next-level Experience (Professionalization of Home Cooking)

If the 'foodie' is the epithome of the change in consumption, then the quest for new experiences matches with the entrance of professional appliances in our households. Kitchen equipment has become one of the targets of manufacturers who employ more and more sophisticated equipment. From the basic to the most complex ones, the evolution in the kitchen make it for more differentiation. Multi-function appliances work by implying the basic sciences and concepts of meal production and food preservation such as mechanical operations, heating, refrigeration and freezing, dehydration, fermentation, acidification, smoking. Rice cookers, slow cookers, electric stock pots, bread-makers or smart stoves, and so-called 'kitchen aid' (multifunction machines which can cut, knead or whip, among basic functions) have all made their appearance in the past decades and have all contributed to relieve

---

[1] The EU has also been actively involved in the field through the PERFORMANCE project (Cordis 2018) which aims to predominately help people that face dysphagia problems and is currently implemented in some 1000 households in Germany already. In America, NASA has been involved and actively engaged in developing the field since 2013 viewing advanced food technologies as the way to tackle space missions' nutritional demands effectively. Though the direct scope of the research is efficiently overcoming the aforementioned problems, greater humanitarian benefits were also stated to constitute indirect powerful drives for the generous funding (Dunbar 2013).

individuals from daily cooking duties. Two appliances introduced some years ago draw more attention. Probably the most famous of the last 50 years is the microwave oven, which has entered homes accompanied by some skepticism; although quite convenient it did not revolutionize home cooking as other gadgetry following it, but interestingly it has undergone a redesign towards a combination of functions allowing new models to have different combinations of cooking techniques (such as grill, steam, classic microwave) and allow for precision cooking. A brand new food technology along this line is ovens using light bulbs through a system initially developed in the solar industry, which reach 500 °C in seconds and without preheating; the innovations is in the technology which allows different and many ingredients, like proteins and vegetables, to be precisely cooked simultaneously (Brava 2019). Given the synergies between different industries of technology, this innovation represents an interesting cross-sectorial innovation; although quite appealing to both highly-skilled cooking-passionate and to no particularly skilled people, its launch price does not make it yet a popular option. Some observers have indicated sous-vide gadgetry as the new appliance which would have changed home cooking following the success in high-end restaurant cuisine. Sous-vide (French for 'under vacuum') has a range of uses for tenderizing textures through gently cooking as the food used is packed in special air-tightened bags (plastic pouches) or glass jars and cooked at low temperatures in a water bath. The resulted food can be consumed right after, although a previous quick searing through pan cooking is recommended, or conserved for some time at low temperature, therefore allowing preparations to be done in advance and make it quite convenient for people with busy schedules. Initially seen in fine dining restaurants, it was adopted by some cuisine lovers using thermal immersion circulators until when some companies have specialized in home equipment, making it a more common domestic equipment. Despite this, its diffusion is not universal. One futuristic innovation is the Sonicprep, a tool which emits ultrasonic sound waves to 'extract, infuse, homogenize, emulsify, suspend, de-gas or even rapidly create barrel-aged flavor' (Sonicprep, 2019). Composed of a generator, converter, probe and sound box, this tool applies low heat vibrations of sound energy, therefore avoiding the transformations given by heat, preserving colors, aromas and nutrients.[2] This one too represents an innovation originally introduced for professionals but then shifted to households. The high price constitutes a clear obstacle for wide adoption despite the appeal for foodies and healthy nutrition followers. Whether the share of consumers willing to buy and, more importantly, able to use these appliances is representative of all consumers is clearly not the case for the moment, but the industry seems to push in this direction.

---

[2] As listed in their website, this homogenizer can make vinaigrettes without using an emulsifier, give wine a fuller and rounder mouth feel, infuse cocktails and other liquids with volatile aromas of fresh herbs or spices, intensify fruit or vegetable pulp for sauces and puree, tenderize and marinate meat in quick time, boost flavor without overcooking fish and other delicate proteins (https://polyscienceculinary.com/products/the-sonicprep-ultrasonic-homogenizer).

7.3.2   Full-Connectivity

In here belong all those technologies which allow for high-end technological equipment both in the kitchen and along the supply chain. The main revolution in this group of technology applications comes from synergies of digitalization and AI technologies; this in fact allows connectivity. We have subdivided this in two groups, one focused on robotization and the other on automated services.

7.3.2.1   Kitchen Roboter and Robotic Chefs

The employment of robots in the catering industry is nothing new, but ultimate models have been employed also in the front-of-the-house operations, as they resemble human beings. Examples are quite common in countries like Japans, where human-shaped robots are now normal to be seen in hotels and also restaurants (Rajesh 2015). The prototypes of these robots mimic human-hand movements with the same efficiency.[3] Robotic chefs have been promised for quite some time apparently. The main incorporation of robotic chefs today happens on the business scale as price barriers hinder wide consumer adoption. Two main examples here: the first is a 'Bionic Bar' released on board of a cruise-ship in 2014 and consisting of a bar's mechanical arm which prepares cocktails ordered using a tablet placed in front of customers. According to the designer's website, the orders begin by "tapping their RFID [Radio Frequency Identification, a tool which uses electromagnetic fields to allow for identification and tracking] bracelet on one of the tablets on display. Besides choosing from standard and signature recipes, guests are able to entirely customize their drink with an almost limitless number of combinations, and have the possibility to personalize it, name their own creation, access their order history and reorder their favorite cocktails, all while rating and commenting on them" (Bionic bar 2018). The designers claims that the drinks that are served will be 'perfect', despite (or arguably because) there is no human involvement in this. The second example is the robotic kitchen developed by a UK-based company[4] and set for consumer releasing in 2019; consisting of two articulated arms, cooking hobs, oven and touchscreen interface, this robot is announced to be able to chop, whisk, stir, pour and clean. The data that guide the production process is being recorded through a multitude of onboard cameras that record human movements.

---

[3] One field-tested addition in the robotic chefs comes in the form of burger-cooking robots (under the anticipating name of 'Flippy'; Miso Robotics 2018). This flipper-robot uses thermal sensors and cameras to get feedback on the grilling process. Consequently, after the CPU evaluates the data, a robotic arm performs corrective adjustments and also serves the burger to the customer research and development already tread the final stages. Investments to implement this technology in the next 2 years in 50 restaurants has been notable (Condliffe 2017).

[4] Moley is a small UK company that through collaboration with Stanford University professors and miscellaneous reputable tech companies like Shadow Robot. The Moley device crosses the threshold of what is considered to be purely 3D food printing and introduces general robotics to the mix.

The patterns are afterwards reproduced by the articulated robotic hands so that each individual, as well as celebrity chefs, can produce and record recipes and meals to upload distribute on the internet (Andrew, 2016). The consumer can select the menu remotely -wirelessly through smartphone apps-, and once the recipe is chosen, pre-portioned ingredients will be delivered at home, so users will only need to place them onto the special containers in the kitchen for robot to begin cooking; it is supplied complete with appliances, cabinetry, safety features, computing and robotics, and fits regular kitchen spaces (Moley 2017). Connectivity with other users will enable bi-directional share of information, recipes and entire cooking patterns.

7.3.2.2 Automated Services

The same combination of technologies we mentioned for the smart kitchen allows also for new services created for consumers. A transition from product-centric to service-centric with focus on consumer is undergoing, driven by the desire for convenience. New routes to the market such as subscription services have become more common, as takeout and ordered food has boomed. For instance, famous sharing-economy transport service Uber has recently launched into food delivering under the name Uber-Eats, and most surprisingly Uber Eats has launched a whole service based on the concept of 'ghost restaurants', virtual restaurants without a full store presence (Tan 2019). Another one is a San Francisco based company specialized in sous-vide immersion circulators, as it has announced their fully connected cooking appliance and, like the final example in the previous section, a service that will deliver frozen, pre-cooked meals which will then be cooked through the same company appliances in brief time. The connected hardware device is in fact tied to a subscription meal service furnished with intelligent auto-reordering system, so once the package has arrived the products are simply scanned thanks to the same device and immersed in a water bath, as the sous-vide gadgetry explained in a previous section. Their website advertised it as a matter of convenience, taste, less food waste and sustainability (nomiku.com 2019). In fact, a section explaining their ethos says "the most delicious food comes from sustainable sources. Those are the people that care about the most holistic ways to feed people. The farmers and butchers we work with are thinking about and acting to create a more sustainable world" (ibid.). Although, strictly speaking, the following is not purely about food consumption but would be classified as service or grocery, a side-note goes in this section to grocery deliveries, as hassle-free grocery deliveries have become a reality. Shopping deliveries have existed for long time, and are now perfected through the means of full automation. In fact, an automated shopping list is also possible as American and Chinese companies such as, respectively, Amazon or Alibaba see the opportunity to fuse home delivery with smart home access control and automatically deliver groceries all the way to the fridge. The experimentation currently undergoing in selected American cities for Amazon (Holt 2019), together with the same company acquisition of natural-food retailer Whole Foods and the creation of unattended and fully automated shops, points at a fully planned change of services and consequently change of consumption habits.

7.3.3   Home-Made Food and Meal Production Modernization

As we said, some technologies are employed at both industrial and household level, making the second at an entry but promising level for adoption of tools which have the potential to redefine much of our relationship with food consumption. Additive manufacturers and 3D food printing are definitive central in this. Although this is covered in details in another chapter in this book, (Chap. 7) here we simply want to address some basics issues that could shed some clarity to make our review. Additive manufacturing of food is being developed by squeezing out food, layer by layer, into three-dimensional objects. A large variety of foods are appropriate candidates, such as chocolate and candy, and flat foods such as crackers, pasta, and pizza. But increasingly this technology is becoming a home appliance, as big machinery are now evolving in little tools to be adopted in the household. For example, a Spanish startup company founded in 2012 has just launched a food printer that can produce snacks comprised of healthy and fresh ingredients quickly, whilst offering complete control over the nutritional value to the consumer. The extruded ingredients are used for surface filling (e.g., pizza or cookie dough and an edible burger from meat paste) and graphical decoration. The company aims at automatizing manufacturing in time-consuming and remote activities (Foodini 2018). This also carries the possibility of creating a preservative-free savoury and crunchy snack on demand at the household level in the near future. This same company considers food printing as the most liable alternative to processed food by overcoming the obstacle of quick meal preparation and cooking, and also a product to reduce food waste thanks to their multiple ingredient capsules already filled with the necessary. This technology has drawn attention specifically when moving to the household segment of the market, for which prices have been lowered consistently.

Lastly, food production: on the front of self-food production many words have been spent. The very notion of backyard-, or rooftop- or community garden dear to the agri-food literature is usually understood as a way to escape mainstream provisioning and regain control over food choices; whether it is permaculture or organic gardening, though, a common theme is growing in spaces outside of the household, to which we can now add technology-based self-production to create food at home, in a hyper-localised fashion. The classic gardening is proposed to be practiced indoor through development of aquaculture and lighting techniques, for which solutions are made available as start-ups (e.g. Urban Leaf; geturbanleaf.com) or crowd-funding producers with prototypes (e.g. AVA Byte 2019; indiegogo.com) spread the word: in the first case with bindle kits made of colored bottles to prevent chemical reactions to light, and a bundle made of seeds, grow lights and accessories comprising soil replacements and germination kits; in the second case buying automated pods, equipped with LED lighting and a smart sensing technology controlled remotely via a dedicated app, and which are soil-free, pesticide-free, self-watering and compostable. In both cases, the drive behind these products, as advertised in respective websites, can be traced back to consciousness about food-related issues, urbanization, nutrition, sustainability, and appeal to like-minded consumers. This particular form of gardening does not require any precedent and specific type of knowledge, and the price of both of them make it affordable to a vast number of consumers.

7.3.4   Enhanced Sensing Through Nanotechnology

Molecular gastronomy has been at the end of last century a brief but explosive example, during which some famous chefs mainly in Western countries have worked together with chemists and physicists to arrive at the transformation of ingredients in unprecedented ways using natural gums, hydrocolloids, nitrogen, dehydrators, enzymes, tools for sferification and other uncommon equipment and techniques. Farrimond (2018) in his article about future food uses the examples of British chef Heston Blumenthal, an exponent of this cooking trend, who had served at the beginning of the millennium a dish called 'Sound of the seas' made of seafood products and first models of mp3 player Apple Ipod to listen to a recording of sounds typical of seas and oceans, such as waves or birds screaming. The proposal was to enhance the enjoyment of that dish. The same chef has also worked on other senses such as sight, supplying customers with 3D glasses when visiting a sweetshop at his indication, as requirements for eating in his restaurant to be taken fully by senses. Farrimond (2018) proposes that senses will be at the forefront of food consumption; he writes that "it is well established that all senses inform the flavor of food: desserts taste creamier if served in a round bowl rather than on a square plate; background hissing or humming makes food taste less sweet; and crisps feel softer if we can't hear them crunching in the mouth. The emerging field of 'neurogastronomy' brings together our latest understanding of neurology and food science and will be a big player in our 2028 dining". Lastly, in this group we want to remind an experiment made at Parisian innovation centre 'Le Laboratoire' in which chefs and chemists worked on encapsulating flavors and developed a way of eating by aerosol, whose first output was an aerosolized chocolate, documented in a fiction book by a Harvard University professor who has then developed other scent additives and also food products (Edwards 2009). The idea behind this one, though, is that some flavor compounds can be combined to enhance the aromatic part of some ingredients through highly specialized technologies (Sensory Cloud, 2019) and creation of specific environments and experiences.

7.3.5   Nutrition and Food Substitution

The final group of food technology innovation comprises nutrition and food substitution. Currently, a branch of (and a huge part of investments on) technological innovation is devoted to biophysical and medicine-related fields. Human nutrition is one of them, for which applications in related food technologies have been tackled extensively also in research (e.g. Dixon 2009). Nutrition is central for understanding the recent trends for both production and processing, as much of the contemporary anxiety for food-related issues can be traced back to it. Research has focused on personalized nutrition based on genetics tests to offer guidance for healthy eating (Farrimond 2018). 'Superfoods', as highlighted in Chap. 6, definitely constitute a major trend. Interest in novel or rediscovered crops indicated as extremely healthy per se (that is, without any need to provide for a balanced or reduced consumption of other less-healthy food, and with no mention of highly-recommended active life) has forced a change in agricultural fields and consequently has driven an adaptation in the global supply chain, to ensure demand could be met. Beside this, in here we focus also on food substitution, as it addresses some social anxiety over ethical issues in nutrition, such as animal eating. Chapter 6 has covered bioprinting as one technology-intensive sector to cover for food demand, with the specificities of in-vitro meat as a meat substitute. But in terms of meat-substitutes, in the past years a full range of plant-based products have emerged. Data tell us that in 2018 in general plant-based food sales rose 20% over the previous year reaching $3.3 billion (PBFA-Nielsen 2018), among which some innovative products stand out. Interestingly cow-milk and cow-milk based products are down, whereas plant-based milk (as it is referred to) gains some positions. Mainly plant-based substitute employ plants because of their characteristics such as less saturated fats, more fibres and Omega-3s, and help with vitamins-income and to reduce blood pressure. The main ingredients are protein-based (such as pea, or wheat, or potato or mung beans proteins) with the addition of some binders such as konjac and xanthan gum, whether it is for egg-substitutes or cooking dough (Ju.st.2019) or for patties to make up a burger.[5] For burgers, the process has started from the flavor of beef burgers and their texture, for which molecules responsible have been searched until the answer was found in so-called heme, an oxygen-carrying molecule present in living plants and animals, later derived only from plants for realizing these burgers. 'Impossiblefoods' website, probably the most pioneering one, explains that for their burger they have been using the heme-containing protein from the roots of soy plants, called soy legume hemoglobin, derived from the DNA of soy plants and then implanted into a genetically engineered yeast, which fermented and then produced more heme. Thus obtained heme are then added to the list of ingredients in their burger, all made from plant-based and vitamin-reach ingredients (impossiblefoods.com 2019). The advocates of these products share an interest in plant-based products driven by concern over animal-based food production for both ethical issues or its consequences on environment and health. These plant-based burgers are available for home consumption and in some eateries, at a quite affordable price.

---

[5] https://www.beyondmeat.com; https://impossiblefoods.com; https://movingmountainsfoods.com

7.4   Food Technology and Consumption Reloaded

As we have found in our research and highlighted in the section on methods, in terms of technologies used and by whom, a clear line is more difficult to be drawn, as many technologies overlap between industrial, catering and household levels (Truninger 2013). This does not mean that these technologies are the same in each sector, as the scale is still significant and it is quite understandable that machinery will be different according to the final destination and user, but many of the innovation of the industry have been and will be adopted elsewhere. Many of these devices are destined to the households, making Lin's (2015) expectations of new interest in home cooking seems correct. The quest to put together edible products from scratch in a mechanized fashion has brought together a mosaic of stakeholders, diverse in every aspect. At the household front, a great deal of competition is taking place to come up with the most ambitious and consumer-friendly device that will enable the forerunner to set foot in this niche - but of high potential - market. Up to now, the food tech field was an industrial solutions mosaic operating mainly on the industrial level and serving high demand food products like readymade meals, snacks and confectionaries. This though has changed heavily in response to the contemporary food trends, but interestingly the different levels of enquiries to which food technology pertains, namely the industrial and the household level, intermingles together with a third level, the one of 'eating out', whether it is as mass consumption or fine-dining. The employment of automated robots and the push for more automated services, as well as the development of tools for self-production, confirms it. For instance, until recently 3D printing has been sugar-based, but technology is emerging that reliably prints savory and fresh ingredients. Historically speaking, it was the American agency for spatial explorations, NASA, that in 2003 declared that they would develop a type of food that could be printed; the main goal of the agency was to ensure that astronauts could print out food, instead of consuming it out of tube, and for this they had to push the boundaries of production, and extend the range of ingredients to be used. In a film fashion, we could say that the predictions common during the 1950s and 1960s of eating capsules or weird products (in movies such as Soylent Green) seems to have been reversed in favor of more complex and tasty meals.

The example offered by 3D food printing is also apt as it pertains to many other experiments in self-production, which reflects the environmental concern and the health-related anxieties of the past 20 years. Nutritious as well as heathy food is at the centre of this. These latter are paramount for food-substitution as well, as much of the justifications employed by advocates of plant-based products employ the same sustainability trope, together with animal rights and ethical eating and living. Soron's (2010) note on green consumption, though, forces us to think that self-production and food substitution might pertain more to a sense of identity and hedonism, as they contribute to social differentiation. 3D food printing allows also to explain more of the professionalization of meal-creation. As technology becomes available in households and prices are lowered, expansion is to be expected. In a consequential mode, this will further push for more professionalization, as a fragmentation of offer is accompanied by further specialization. Professionalization here means resembling the industry and the cooking professionals, as this figure embodies the new fashion for the media industry. Specialization and precision, both in the home-context and for hospitality and catering, will be pushed even further. The chances offered by more precise cooking, in which each operation can be further subdivided and defined, will open further spaces, as was the case for molecular gastronomy. A push on senses and sensory use is what characterizes this category. Increasingly, the industry has specialized in the natural compounds present in food products, their molecules and chemical composition to work on infinite combination that would give the industry more resources to work on. The food processing industry has employed and developed these techniques for decades now, as a matter of concentrating or enhancing flavors to provide for better and more mouth-watering products. Sensoring is promised to push the boundaries of our sensible knowledge of the world, as Farrimond (2018) has commented. The tendencies of globalization and aestheticization Warde (2015) mentions seems to be at play in many ways. The global phenomenon of celebrity chefs, as well as travel- and food- television shows and magazine articles, or the emerging figure of the bloggers, is testimony to this. Nice-looking dishes in which healthy as well as costly, inexpensive, traditional, authentic or ethnic food are displayed, are a consequence of this exteriorization. It is then correct to individuate foodies as the symbol of this tendency, as the foodie is probably the person who can embody transversally the main categories we have discussed before.

The main revolution of technology applications comes from synergies of digitalization and AI technologies, as it has been highlighted when talking about full connectivity. This allows interaction of equipment, voice assistants and chatbots helping with the cooking process. This presupposes that adopters know how to use these technologies, which have been rendered more accessible on the basis of the 'user-friendly' imperative rule. The outcome makes an observer think at a digital kitchen in which little human participation is necessary, as the instruction set for our appliances make the content becoming dynamic, atomized and personalized depending on our personal preferences and the context of our current day, meal plan, and food inventory. This has an impact on food-related literacy and knowledge, which seems to become redundant and in the hands of few. On the other side, though, users are able, and encouraged, to share their own cooking recipes and patterns (think about, for instance, the upper end of robotization that can be assumed in the introduction of computer chefs who mimic human movements to prepare meals, as human-hand patterns are efficiently reproduced and performed by employing pre-configured motion libraries that govern the mimicking of human movements like picking up, putting down or pouring) over the internet or download creations by celebrity chefs or other plain users (Andrew 2016). This would contradict Grignon's (1996) idea of further individualization, although it can be observed that the types of interaction are not spontaneous but always regulated through external factors independent of a specific context. Of course, and as a final note, the massive use of apps and related technology means that data are collected and used by the industry behind it, raising questions over use of the same. This does not seem to bother users; the issues of data ownership that activists in the tech field are so vocal about, though, will definitely become reason for concern in this field too.

7.5   Conclusions

Consumption patterns are constantly changing, and arguably they will change even more in the coming decades. Food consumption changes according also to the technologies available, but at the same time it is clear that technologies themselves have changed according to the use and demand of consumers. The Science and Technology Studies we have mentioned elsewhere in this book are definitely able to provide more insights on the interaction between humans and non-humans, but this would exceed the defined boundaries of this chapter. After a scrutiny, it seems that much of the innovations in the way we eat and interact will come from factors outside of basic food products, as technology will have taken care of the preparation operations and, for both eating out or consuming food in our households, the next steps drive us toward immaterial, not tangible but technology-enabled landscapes. These of course are changing the way we interact and relate to each other, in both household and outdoor consumption; the example of apps that can act as a medium for sharing recipes or create a forum of course does not signal clear interaction or closeness other than a common interest; on the other hand, full automation can increase distanciation between final consumers and meal-producers, for instance, as there is no room for interaction. We can imagine, as many tech advocates do, full connectivity, an advanced internet of things to the point that it will be enough to just tap on our smartphones for organizing the week-meals of our children and dear ones from remote and maybe even distant places, which can be of great help for working mothers (mainly) and fathers or care-takers, but of course we should also ask what will be lost in terms of personal relationships, as the future envisioned by tech moguls is not unfolding in that precise direction. The trends in ethical eating, health concern, environmental issues are at the forefront, but adoption of new technologies are to be measured against disposable income and education/food and tech literacy, two of the main barriers common to many of the categories proposed. Although prices have dropped in many cases and is usually taken as a good sign for market development (or as some like to say, for democratization of consumption), it is yet to be confirmed that adoption will be immediate. Likewise, cooking as a family-caring or recreational activity is understood and experienced differently by different individuals, who might find the role of technology as foundational or intrusive on the basis of their relationship with it; where, then, is the line to be drawn to understand when has technology pushed too much? Should the market be the judge in this or we risk losing something out of full technologization of food-related activities? The same of course can be said for food consumed outside of the household, where a certain level of craftsmanship is still preserved as a marker of differentiation (think about high-end restaurants that can charge extremely high prices for a meal consumed there, on the basis of artisan preparation and of a level of unique experience). Further, a clear line cannot be drawn easily on matters of 'naturalness', so dear to both environment- and health- concerned consumers, as the applications of further technologies complicates the debate: to what extent is a plant-based food substitute more 'natural'? Of course to answer this question it is necessary first of all to define what is 'natural', a task we will reserve for future research. Ethical issues in terms of animal welfare and rights are superseded here, but what sort of other consequences can come from allowing this kind of genetic research and application needs serious debates. Lastly, at the core of sociological research, although these technologies seem to make no specific difference in terms of gender, ethnicity or religion, these continue to play a role in terms of food choices and of task distribution (or labour division, if you prefer), whether it is in the household or in a professional environment. Despite the promises of technology to overcome the barriers provided by education or gender, hindrances will remain, as they are more profound that technological optimism might hope (see also Chap. 8 for social stratification consideration on consumption and adoption of new techs). So far, then, a final world cannot be spent, as it has to do with factors outside of the specific field of food consumption (namely, how much we perceive technology to be neutral or value free).

As a final consideration which could not be addressed in this chapter as it would exceed the scope of it, what will need to be addressed in the near future and is just at the beginning of societal concern is the implications that the use of modern technologies based on data collection will have on food consumption. As highlighted in other chapters in this book, data are valuable currency, and the pie is so large that no big player in the field will take a step back unless clear boundaries will be institutionally set.

References

Andrew, E. (2016). Robot chef that can cook 2,000 meals set to go on sale In 2017. Resource document. http://www.iflscience.com/technology/robot-chef-home-could-arrive-2017/. Accessed 10 Feb 2018.

AVA Byte (2019) Resource document. AVA Byte: Smart, Simple, Sustainable Indoor Garden. https://www.indiegogo.com/projects/ava-byte-smart-simple-sustainable-indoor-garden#/. Accessed 23 Jan 2019.

Bionic Bar. (2018). https://carloratti.com/project/bionic-bar. Accessed 15 May 2018.

Brava. (2019). Brava. Available at: https://www.brava.com/products/marketplace/brava/oven/. Accessed 8 Dec 2018.

Campbell, C. (1983). Romanticism and the consumer ethic: Intimations of a weber-style thesis. *Sociological Analysis, 44*(4), 279–296.

Carolan, M. (2012). *The sociology of food and agriculture*. London: Routledge.

Corrigan, P. (1997). The sociology of consumption: An introduction. Sage.

Cordis. (2018). European Commission: CORDIS: Projects and Results: Development of Personalised Food using Rapid Manufacturing for the Nutrition of elderly Consumers. retrieved February 11, 2018 from https://cordis.europa.eu/project/rcn/105482_en.html

Condliffe, J. (2017). Robotic chefs are getting a little better (as long as you like fast food). https://www.technologyreview.com/s/603830/robotic-chefs-are-getting-better-if-you-like-fast-food/

Diner, H. (2001). *Hungering for America: Italian, Irish and Jewish Foodways in the age of migration*. Cambridge, MA: Harvard Univ. Press.

Dixon, J. (2009). From the imperial to the empty calorie: how nutrition relations underpin food regime transitions. Agriculture and Human Values, 26 (4):321-333.

Dunbar, B. (2013). 3D Printing: Food in Space. retrieved February 5, 2018 from https://www.nasa.gov/directorates/spacetech/home/feature_3d_food.html

Edwards, D. (2009). *Whiff*. Cambridge, MA: Harvard University Press.

Farrimond, S. (2018). The future of food: what we'll eat in 2018. Resource document. https://www.sciencefocus.com/future-technology/the-future-of-food-what-well-eat-in-2028/?utm_campaign=The+future+of+food%3A+what+we%E2%80%99ll+eat+in%C2%A02028&utm_medium=referral&utm_source=AppleNews. Accessed 11 Nov 2018.

Foodini. (2018). https://www.naturalmachines.com/press-kit/. Accessed 14 May 2018

Germov, J., & Williams, L. (Eds.). (2010). *A sociology of food and nutrition: The social appetite*. Victoria: Oxford Unviersity Press.

Goodman, D., & DuPuis, E. M. (2002). Knowing food and growing food: Beyond the production consumption debate in the sociology of agriculture. *Sociologia Ruralis, 42*(1), 5–22.

Grignon, C. (1996). Rule, fashion, work: The social genesis of the contemporary French patterns of meals. *Food and foodways, 6*(3), 205–241.

Geturbanleaf.com. (2019). Urban Leaf | Gardening Gifts & Beginner Indoor Herb Gardens. Resource document [online]. https://www.geturbanleaf.com. Accessed 11 Jan 2019.

Harrington, J., Fitzgerald, A. P., Layte, R., Lutomski, J., Molcho, M., & Perry, I. J. (2011). Sociodemographic, health and lifestyle predictors of poor diets. *Public Health Nutrition, 14*(12), 2166–2175.

Holm, L. (2013). Sociology of food consumption. In A. Murcott, W. Belasco, & P. Jackson (Eds.), *The handbook of food research* (pp. 324–337). London: Bloomsbury.

Holt, K. (2019). Amazon starts testing its 'scout' delivery robot. Resource document https://www.engadget.com/2019/01/23/amazon-scout-delivery-robot/?guccounter=1&guce_referrer_us=aHR0cHM6Ly9kdWNrZHVja2dvLmNvbS8S8&guce_referrer_cs=1ELSwfE1J-jrnWXO5lVx-A. Accessed 28 Jan 2019.

Impossiblefoods.com. (2019). Resource document [online] https://impossiblefoods.com. Lastly accessed 26 Oct 2018.

Ju.st. (2019). JUST. Resource document. [online] https://www.ju.st/en-us. Accessed 12 Sept 2018.

Lin, C. (2015). 3D food printing: A taste of the future. *Journal of Food Science Education, 14*(3), 86–87. https://doi.org/10.1111/1541-4329.12061.

Lupton, D. (1996). *Food, the body and the self*. London: Sage.

Lee, D. & Newby, H. (1983) The problem of Sociology. London: Hutchinson.

Mennell, S., Murcott, A., & Van Otterloo, A. H. (1992). *The sociology of food: Eating, diet and culture*. London: Sage Publications.

Miso Robotics. (2018). Miso Robotics | The Future of Food is Here. [online] Available at: https://misorobotics.com/. Accessed 12 Oct 2018.

Moley (2017). Moley – The world's first robotic kitchen. [online] http://www.moley.com/. Accessed 15 Dec 2017

Nomiku.com. (2019). Nomiku Meals: Restaurant worthy meals with a wave of your hand. [online] http://www.nomiku.com. Lastly accessed 12 Dec 2018.

Patel, R. (2013). The long green revolution. *The Journal of Peasant Studies, 40*(1), 1–63.

Paterson, M. (2006). *Consumption and everyday life*. London: Routledge.

PBFA-Nielsen. (2018). Resource document. https://plantbasedfoods.org/wp-content/uploads/2018/07/PBFA-Release-on-Nielsen-Data-7.30.18.pdf. Accessed 11 Dec 2018.

Rajesh, M. (2015). Inside Japan's first robot-staffed hotel. https://www.theguardian.com/travel/2015/aug/14/japan-henn-na-hotel-staffed-by-robots. Accessed 13 Sept 2018.

Sassatelli, R. (2007). *Consumer culture: History, theory and politics*. Oxford: Berg.

Schor, J. B. (2007). In defense of consumer critique: Revisiting the consumption debates of the twentieth century. *The Annals of the American Academy of Political and Social Science, 611*(1), 16–30.

Sensory Cloud. (2019). sensory cloud. Resource document [online] https://sensory-cloud.com. Accessed 18 Jan 2019.

Sonicprep. (2019). The Sonicprep™ Ultrasonic Homogenizer. [online] PolyScience Culinary. https://polyscienceculinary.com/products/the-sonicprep-ultrasonic-homogenizer.   Lastly accessed 19 Nov 2019.

Soron, D. (2010). Sustainability, self-identity and the sociology of consumption. *Sustainable development, 18*(3), 172–181.

Thompson, C. J. (1996). Caring consumers: Gendered consumption meanings and the juggling lifestyle. *Journal of Consumer Research, 22*(4), 388–407.

Thompson, K. (2016). Sociological theories of consumerism and consumption. https://revisesociology.com/2016/10/12/sociological-theories-of-consumerism-and-consumption/. Accessed 15 Sept 2018.

Truninger, M. (2013). The historical development of industrial and domestic food technologies. In A. Murcott, W. Belasco, & P. Jackson (Eds.), *The handbook of food research* (pp. 82–108). London: Bloomsbury.

Tan, Y. (2019). UberEats is going to let you order food from 'virtual restaurants' that don't exist IRL. [online]   Mashable.   https://mashable.com/2017/11/09/ubereats-virtual-restaurants/?europe=true#3rju31J_Smqx. Accessed 15 Jan 2019.

Veblen, T. (1994). *The theory of the leisure class*. Harmondsworth: Penguin.

Warde, A. (2015). The sociology of consumption: Its recent development. *Annual Review of Sociology, 41*, 117–134.

Warde, A., & Martens, L. (2000). *Eating out: Social differentiation, consumption and pleasure*. Cambridge, UK: Cambridge Uniersity Press.

Warde, A., Cheng, S. L., Olsen, W., & Southerton, D. (2007). Changes in the practice of eating. *Acta Sociologica, 50*(4), 363–385.

Zelizer, V. (2005). Culture and consumption. In N. J. Smelser & R. Swedberg (Eds.), *The handbook of economic sociology* (pp. 331–354). Princeton: Princeton University Press.

Chapter 8
Technologies at the Crossroads of Food Security and Migration

Lubana Al-Sayed

8.1 Introduction

One of the arguments that paved the way for the development of food industry was to enhance the food security situation at individual, societal, national, and even international levels. This justification was used to improve their social legitimization and facilitate their proliferation. However, when looking at recent figures published by the World Health Organization (WHO), indicating that 1.9 billion adults are overweight or obese and 462 million are underweight (WHO 2018), one starts questioning the legitimacy of the food industry and the promised role of future advancement in food technology and food digitalization. How is the industry going to contribute to driving changes in the current situation towards a healthier individual and a healthier planet? This question is highly relevant when thinking about the negative impacts of the capitalist economic relations around the globe, which has resulted in severe inequalities among nations, regions within a nation, classes, and different ethnic groups. These inequalities represent a considerable burden that prevents our economies from delivering universal health, education and other public services, according to the Oxfam Report 2019 (Lawson et al. 2019). Globalization, urbanization, and industrialization have shaped new lifestyles, new tastes and consumption patterns, and new forms of inequalities which affect the conditions of food provisioning and food accessibility. Larger segments of the world population have become alienated from the means of production and have lost skills on how to grow their own food and to subsist themselves. Hence, they turn out to be dependent on the availability and accessibility of food on the market. Thus, providing cheap food becomes crucial for capital (the accumulation process, as Marxist theorists would remind us) and for achieving political stability. Consequently, the low-income

L. Al-Sayed (✉)
Societal Transition and Agriculture, University of Hohenheim, Schloss Museumsfluegel, Stuttgart, Germany
e-mail: lubana.alsayed@uni-hohenheim.de

© Springer Nature Switzerland AG 2019
C. Piatti et al. (eds.), *Food Tech Transitions*,
https://doi.org/10.1007/978-3-030-21059-5_8

population faces extreme difficulties in accessing healthy food. They mainly depend on cheap processed food to allay their hunger, which risks leading to serious health problems, such as obesity, diabetes, and cardiovascular diseases. With the increased political instability that we have witnessed recently and the rise of new conflicts, poverty, displacement, and forced migration, this segment of the population (low-income population) is proliferating and the already existing gap between rich and poor gets wider and wider. A larger stratum of society becomes dependent on food aid programs, such as food stamps in the US, and food banks in Canada and Germany. Many studies have shown the high prevalence of food insecurity among the beneficiaries of food aid programs (Depa et al. 2018; Fowokan et al. 2018; Simmet et al. 2018; Dinour et al. 2007). Migrants represent an important share of those who benefit from food aid programs, particularly in these times where migration flows have escalated. In 2015 and 2016, migration has become a recurrent topic (not only relevant to policymakers and social workers) and perceived as an emergency and political issue. Yet, there are good reasons to consider immigration more than just a temporary trend. According to IMO (2017), over one billion people are migrants, mainly as a result of changing geophysical and geopolitical situations, which was not unexpected. Just to name one, already in 2009 Tim Jackson warned us that mass migration combined with climate change and related wars would be the concern of the new generations (Jackson 2009). Therefore, it is critical to understand if and how the food industry is going to provide support to tackle the increased social inequality in this era of strong migration flows. This is a highly relevant question, not least since one of the promises made by the advocates of food technology is to make life better for all, and not just for a narrow segment of the society, who usually are people with some form of 'capital'. In this chapter, I will first shed light on the alleged role of food technology in promoting food and nutrition security, and I will critically analyze the promises of the new food technology and food digitalization in improving the current situation. Then, since we are now living in an era of strong migration flows, and because migrants in developed countries usually face multiple food-related challenges, such as forced changes in dietary habits and little knowledge on how to use resources available, I will highlight the link between food security and migration, analyzing the role of food technology and, to open up a bit the terrain of discussion, food-related technology in improving the life of migrants, as these have become part of our relationship with food. Finally, I will use as a case study the Syrian forced migration to Germany to show the transition in their food-related habits and their interaction with the new food-environment and the role played by new technologies in enhancing or undermining their food security situation. More precisely, I will focus on how refugees are making use of modern technologies food-related in a participatory way, which aims at empowering them and easing their food-related life in the new food environment.

8.2 Food Technology as Enhancer or Inhibitor of Food Security?

Humans have developed food processing operations since about 2 million years ago (Wrangham 2009) when our ancestors discovered cooking, which is considered the main form of food processing. Later, human beings developed other technologies, such as fermentation, drying, and salt preservation techniques. All of these practices have developed with the aim of transferring perishable food to a more durable one, and to increase the food availability throughout the year regardless of seasonality. Progressively, the food processing operations became more and more complex and sophisticated with the objectives of producing abundant, varied, convenient, and less costly food with the ultimate goal of promoting the food security situation and feeding the rapidly growing numbers of the population (Floros et al. 2010). Still, nowadays, despite the vast improvements in both technology and developmental programs for ensuring food security, 815 million people go hungry (FAO 2017). So, to what extent have food technologies promoted food security? The concept of food security has gone through multiple phases. Its conceptual roots emerged in the aftermath of World War II, where the fighting countries were suffering from a severe food shortage. During that time, US president Roosevelt in 1941 advocated for four essential freedoms that tie all human together: freedom of speech, of worship, from want, and from fear. Then the Food and Agricultural Organization (FAO) has embarked on the freedom from want in defining the principle of food security. FAO started to advocate for promoting the human ability to obtain food in a decent way and to achieve freedom from want, which is accomplished not only by reaching the "security of food", but also the "security through food" (Carolan 2013, p. 36). The spirit of this concept was considerably strong at that time. However, later when translating the notion of food security into practice, the practitioners and the policy-makers have reduced its meaning to a rather narrow concept of productivism, which focuses on producing enough quantities without considering the social and environmental impacts of such production. The policies and strategies that defined conditions of food provisioning have paved the way to mechanizing the food system, with the aim of increasing the level of specialization, simplification, and standardization of the entire food chain. By achieving that, the industry has increased the efficiency and reduced the cost of production, what is known as "economies of scale" (Carolan 2013). The evolution of nutrition science in the 1940s, particularly the adoption of the calorie as an effective metric for human energy requirements and as a unit to measure food, has shifted the understanding of food security through the lens of calorie (Dixon 2009), what Michael Carolan called "the calorie-zation of food security" (Carolan 2013, p. 27). Accordingly, the policies and strategies for enhancing food security were mainly based on producing enough calories, regardless of their types and the way of producing them, to sustain the growing population.

Progressively, the advancement in nutrition science, the discoveries of macro- and micro-nutrients, and food technologies have provided the food industry with a broader platform for profit through 'de-nutrifying' and 're-nutrifying' food supplies. Thus, the food industry based on science in differentiating their products by adding more value (in forms of nutritional claims; Dixon 2009). By focusing on calories and prices (enumerating food), people's food-related knowledge became impersonal and distanced from any cultural, social, and geographical influences. It has shifted away from taste and experience (traditional food-related knowledge) towards calculations, which somehow still resonates with us (Dixon 2009; Mudry 2006). Later, from the 1970s until now, the governmental policies and the corporate strategies have focused on trade liberalization, the valorization of comparative advantage, and the integration of the global market; the fruits of neoliberal reform. This reform has had a negative impact on food accessibility where it reinforced the production of cheap calories and created inadequate access to food, what Carolan (2013, p. 38) called "the neo-liberalization of food security". During that time, the discourses of food security have shifted from rights-based language towards a market-oriented one that commodified food and shifted the food security objectives from the national level to household level (Koc 2013). Following market liberalization, there was a proclivity towards finance liberalization (circa from the 1980s on), which has led to the proliferation of Foreign Direct Investment (FDI). Big corporations have started to produce processed food, which is nutritiously shallow but energy dense (what has been identified as an empty calorie), around the globe. Consequently, the fast food industry has entered new markets spreading the culture of processed food and controlling over the national food system (Carolan 2013; Dixon 2009).

The prevalence of a culture of processed food has led to serious limits on the range of real choices available to an average person seeking to express his or her identity and values through purchases (Jaffe and Gertler 2006), as well as to a great inequality between the different strata of the society, not to mention the severe environmental degradation. In this context, the poorer population's diet is restricted to cheap, highly processed and energy dense food products, while affluent people enjoy a variety of foods, and increasingly expensive fruits and vegetables. The shift in diet and the sedentary lifestyle have led to the spread of obesity and diet-related diseases, a so-called "nutrition crisis" (Dixon 2009, p. 322). Likewise, people are systematically deprived of the information and knowledge to make informed food choices, what Jaffe and Gertler (2006) have termed consumer deskilling inspired by Harry Braverman's theory of workers deskilling (Braverman 1998). A more significant segment of the population became increasingly distanced from the production sites and processes. Food manufacturers took advantage of this point, increasingly spending millions on marketing campaigns to re-educate consumers for their own purposes while pretending to respond to public demand (Carolan 2017; Dixon 2009).

The changes in lifestyle and flexible working conditions, in terms of time (overtime, shifts, on-call, commuting, etc.), have created "flexible consumers" (Jaffe and Gertler 2006, p. 145). A lot of individuals started to eat at odd hours, on the run, and mostly alone, which has reinforced the dependence on processed food, such as take-aways, fast-food, and ready meals. Thus, manufactured food has significantly replaced the notion of cooking from scratch. Food industries have established to produce ready-to-eat food and snacks. Food has become increasingly subject to scientific and industrial processes, and the journey from the field to the table has become further complex. Consequently, the gap between what consumers know and what actually happens in practical and behavioral terms has widened. This gap has led to "a growing gap in power, and a growing capacity for manufacturers and retailers to manipulate taste and buying behavior" (Jaffe and Gertler 2006, p. 145). In the last decade, there was an urge to change this situation due to the high economic burden of malnutrition. Governments started to influence food-buying behavior by regulating the prices of certain food commodities through taxation and subsidization. For instance, in the United States and Canada, a taxation on high energy dense food was applied as an approach to limit their consumption (Swinburn and Egger 2002). On the other hand, scientists have advocated reclaiming the original spirit of food security (see for instance Brown 2012; Carolan 2013). Even FAO has taken a decision to incorporate the concept of food sovereignty into food security as a way to rethink the meaning of and to promote and achieve food security (Lawrence and McMichael 2012). Under this pressure, food technologists started to shift their focus from maximizing the efficiency of the production process and reducing costs as much as possible, towards adopting new innovative techniques to produce constantly more diverse, safe, healthy, and environmentally sustainable products.

New emerging technologies have been developed to produce fresh-like, 'minimally' processed, highly nutritious, and safe food, such as novel thermal processes (microwave and ohmic heat), and non-thermal physical preservation methods (high-pressure processing (HPP), pulsed electric fields, ultrasonic waves, high-intensity pulsed light, and others, see Chaps. 1 and 2). These novel technologies form a potential interest in the industry. Though their current application is still limited (to a less extent thermal processes and HPP), the main reason behind the late commercial spread of these technologies is the high cost of their equipment (Butz and Tauscher 2002). Moreover, the development of nanotechnology, food digitalization, and additive manufacturing, to which the best example is 3D food printing (see Chaps. 6 and 7 in this book), are now considered to have a high potential to tackle malnutrition through the customization and personalization of the diet, all of this while retaining costs low (Chadwick 2017). Through the usage of smartphone applications (apps), the industry is able to deliver adequate nutritional information according to individual's requirements (age, gender, and specific dietary requirements). Previously, the focus of major food companies with their apps was on games and entertainment and on connecting consumers to places where they can find their products (e.g., Coca-Cola Freestyle). However, now the industry has started to shift towards creating fully personalized and integrated nutrition apps, which guides consumers to make healthier food choices (e.g., Pediatric Nutrition Calculator, and MyTubefeeding applications designed by Nestle). On-demand production and reproduction of traditional foods, such as pizza, using new, not widely accepted, but more nutritious ingredients, such as insects and algae, are considered essential to promote food and nutrition security situation. The proponents of this technology insist on its ability to address resource constraints, reduce waste, and produce healthy and reduced ecological footprint food, thus it could be a way to address climate change, and food insecurity in times of natural disasters (Nair 2016). Still, the advocates for the use of emerging technologies value similar concepts, such as productivity, convenience, readily available, and cheap food (see FMI and Kurt Salmon 2017). Furthermore, the introduction of new technologies always comes with a new range of skills and values which in turn erase previously practiced skills (Jaffe and Gertler 2006). For example, when bread-maker machines were developed, people started to adopt them, and some began to make bread at home in a traditional manner (cooking from scratch), while others relied on the pre-bread mixes produced by the food industry which mimics the ingredients and the taste of processed bread (Jaffe and Gertler 2006). Similarly, the adoption of a 3D food printer as a kitchen appliance will make significant strata of the society to be dependent on the pre-filled food capsules produced by the new food industries since they are faster and more convenient. Meanwhile, the content of these capsules is puree-like, and thus it is difficult to recognize their ingredients, therefore consumers will most likely become increasingly distanced from what they eat, and gradually they will lose their previous skills which will be hard to get back, since "technology, deskilling, and new food products co-evolve in a dialectical fashion" (Jaffe and Gertler 2006, p. 147). The development and the adoption of a new kitchen appliance generate new opportunities for the learning of new skills and creating new products, whilst at the same time replacing previously held skills by a machine. Accordingly, the adoption of new technologies requires re-educating people on how to handle them and to use them in the best ways to make healthy and sustainable food choices. With the proliferation of mobile phones and tablets, new software applications have been developed to inform and facilitate the process of food choices with the aim of promoting health (e.g. Fooducate app). The development of food digitalization and Internet of Things (IoT) simplified the process of data collection, such as collecting information concerning the assessment of the dietary intake, which is now used to provide personalized advice and suggest diet and nutrition-related decisions. These applications could have a high potential in modifying unhealthy habits for a segment of the population (West et al. 2017; Franco et al. 2016). Although these applications contain an extensive database and rely on a barcode system, assessing the dietary intake of a traditional diet is still complicated and manually inserting the nutritional information requires a high level of nutritional knowledge. Besides, culturally appropriate health messages and feedback, especially for individuals with low health literacy are still lacking (Coughlin et al. 2015). Therefore, these applications are currently targeting a specific segment of the population, who are educated, relatively young, with good mobile literacy, and higher income (Carroll et al. 2017; Mackert et al. 2016). This situation is more aggravated in developing countries, where the utilization of food and nutrition application is very limited, and the process of assessing dietary intake is much more complicated under the absence of a comprehensive nutritional database. However, are not all these modern technologies targeted at a specific category of society (socially patterned)? How is the industry going to tackle the social inequality in light of the increasing gap between rich and poor especially in this era of strong migration flows? These questions are pivotal since the notion of cheap food, and Engel's law (which proposes that when household incomes rise, the percentage of income allocated to food decreases, see Engel 1857) seems to be coming to an end in both developed and developing countries, and thus the low-income populace will spend most of their income on purchasing food rather than investing in obtaining new fancy technologies, such as 3D food printer or smart oven.

8.3 Migration and Food Security

In this context of relative food security and quest for more sustainable and healthy food, migration adds another level of consideration and concern. We are currently living in a new era of migration where, according to the International Organization for Migration (IMO 2017), over one billion people are migrants. According to recent global estimates, there were around 244 million international migrants in the world in 2015, which equates 3.3% of the global population, and about 740 million internal migrants, in addition to 68.5 million (UNHCR 2018) people forcibly displaced due to conflicts and human rights violations. The factors that push people to flee their countries of origin and the ones that pull them towards their chosen destination are diverse and multifaceted. According to a recent report on food and migration (MacroGeo and BCFN 2017), food systems are part of these push-and-pull factors. Indeed, recent major migratory movement have resulted from a turmoil in the traditional food systems, due to "climate change and droughts (Sahelian countries in the 1970s), inadequate food policies (Ethiopia in the 1980s), controversial trade agreements (West African countries since 1990s)," or armed conflicts and their implication on food availability and accessibility (Syria, Yemen, Iraq, Libya in the 2010s). In the destination countries, usually the developed ones, the lack of labor in agro-food sectors has acted as a pull factor for those migrants and facilitated the exploitation of inexpensive workers (MacroGeo and BCFN 2017). The link between migration and food is much more profound than one might think. After arriving in a new destination, migrants have to adapt to different lifestyles and are confronted with varieties of economic and social adversities, which results in psychological discomfort and stress. This process is usually much more challenging for asylum seekers and refugees due to a great deal of uncertainty they experience and the post-traumatic stress and emotional problems resulting from loss of family and social support (Rosenblum and Tichenor 2012; Carswell et al. 2011). Thus, the process of integrating them in the host society is much more challenging. And, as highlighted recently, food has turned out to be a central factor in assisting migrants to settle and be integrated into the host communities (MacroGeo and BCFN 2017).

Food plays symbolic and hedonic goals and becomes a distinctive element of the identity of individuals and communities. Migration processes represent possibilities possibilities to change and resist to novel habits, different behavior and new cultural experiences (Koc and Welsh 2002). It is thoroughly documented that many migrants exhibit a better health status than the resident population. This phenomenon is known as "healthy migrant effect" (Fennelly 2007). However, this situation quickly deteriorates with the changes in the new environment. Migrants from developing countries start to experience tremendous shifts in their food-related practices (e.g. purchases, preparation, budget management) after moving to developed countries. They acquire and adjust certain aspects of their food-related habits affected by the customs of the host society in a process called "dietary acculturation" (Lesser et al. 2014; Hassan and Hekmat 2012; Dharod et al. 2011; Hadley et al. 2007; Barry and Garner 2001). The acculturation process might have a severe impact on the food security situation of migrants through several pathways. They might seek traditional food items (commonly consumed food in their country of origin) which are most of the time expensive and not compatible with their current monetary situation (Hadley et al. 2007; Jetter and Cassady 2006; Drewnowski and Specter, 2004). Migrants might face problems with home budget management, especially when they send remittance to their family back in the country of origin, which could lead to the early completion of income (Pérez-Escamilla et al. 2000). Besides, the difficulties to adapt to the new food environment and language barriers may lead to severe food illiteracy, make the food choice process more complicated, and undermines their ability to access, understand, process, and use basic food and nutrition information. Thus, the adaptation to the new food environment and language barriers might be coupled with the aforementioned challenges and exacerbates the food insecurity situation of migrants (Hadley et al. 2007, 2010). Acculturation-related changes may increase the risk of obesity and diet-related diseases as a result of the adoption of unhealthy dietary patterns and sedentary lifestyle (BURNS 2004; Hersey et al. 2001; Holmboe-Ottesen and Wandel 2012; Patil et al. 2010). The adoption of unhealthy diets results from comparatively lower incomes than host society, limited social capital, and inadequate dietary intake which is characterized by reduced food intake, lowering diversity, and increased consumption of energy-dense food (Hadley et al. 2007). The food environment in developed countries is loaded with convenience food, and since migrants have to be integrated into the host society and to build their lives again from zero, they most often lack time to prepare their food the way they used to in their home countries. The social and economic pressures have a detrimental effect on their relationship to food. They begin to rely more on convenience, affordable foods which do not clash with their religious and cultural beliefs (Oussedik 2012). Thus, gradually their traditional food knowledge fades with time (Geissler and Powers 2010; Kwik 2008). Therefore, the food security situation of migrants is considered a determinant of both their physical health and their overall well-being.

Accordingly, what is the role of food technology in here? A crucial question is whether and how the food industry will be able to make available better options for fresh foods that are respectful of migrants' traditions and needs, enabling them to express their identities, whilst also being affordable for the majority of migrants who typically suffer from significant economic pressures, hence decreasing current social inequalities. If we think about the minimally processed food items produced using emerging technologies, they are likely to be more expensive than regular food products. Thus, their consumption will reflect the already existent divide between the different strata of the society and will reinforce much of the inequalities. The low-income population, which includes migrants, will not be able to afford these types of products. Besides, they will not be able to afford the adoption of innovative food appliances, such as 3D printers, robo-chef, or smart oven in the short and medium run. In the long run, when the prices of highly modern and fancy appliances such as food printers go down, low-income population might afford it, but since the food prices will arguably continue to increase and the Engel's law aforementioned is expected to reverse (Rosin et al. 2013), impoverished people will most likely spend most of their income on purchasing basic food to feed themselves. It is worth mentioning that there are fundamental hurdles in the adoption of new food technologies in developing countries. The technological adoption will likely reflect and reinforce the already existent divide between developed and developing countries, since the readiness for welcoming IoT technologies is still absent in developing countries, due to political instability, economic and governance problems, and connectivity issues. The technological development usually happens according to multiple geopolitical factors (Bey 2016). What is most crucial for migrants in developed countries is to enhance their ability to navigate the food system properly and to foster their feelings of belonging and welcoming, since food security for those people involves a feeling "at home" (Koc and Welsh 2002). Ensuring that migrants have skills, knowledge, and resources to navigate their new food environment will enhance their food and nutrition security and thus their well-being. The migration process does not only change the eating habits of migrants, but it also has an impact on the food environment in the host society, which with a degree of reticence influence the eating habits of the host community. Importing new ingredients, opening new small supermarkets, ethnic restaurants, and fast food outlet (such as doner kebab) have proliferated recently with the increase in the number of migrants. This new food environment supplies the public with new food items, rules, and cuisine (MacroGeo and BCFN 2017; Oussedik 2012).

8.4 Exploring Experiences of Changing Food Environment Among Syrian Refugees Living in Stuttgart, Germany

Food and eating practices influence and are influenced by miscellaneous determinants, such as social, economic, cultural, geographical, physical, psychological, environmental, and political factors (Counihan and Esterik 2013). As already documented, food plays a vital role in expressing cultural and ethnic identity. During migration,[1] this symbolic function becomes much more essential for migrants, and has a significant effect on their food choices to enable them to find a bridge between

---

[1] In this section, we focus on forced migrants, hence from now on with term 'migrant(s)' or 'migration' we will mainly refer to forced ones.

their pre- and post-migration life (Koc and Welsh 2002). The Syrian migration constitutes one of the most significant migratory flows in recent decades, and therefore provides a meaningful case study when exploring the intersection of food security and migration. The conflict in Syria started on March 2011, and until now, tragically, there is no clear prospect of resolution. This conflict has led to a vast forced mass migration, where 6.1 million are displaced inside Syria (IDPs), 5.3 million refugees in the neighboring countries, in addition to 970,000 refugees having applied for asylum in Europe (HNO 2018). Germany has received and hosted the largest share of Syrian requests for asylum in Europe (Eurostat 2018). Therefore, the Syrian refugees and asylum seekers interaction with the new (German) food environment is worth exploring as a case study to navigate some of the real-life implications of changing the food environment in transnational prospect. In-depth semi-structured interviews have been conducted with 34 Syrian refugees and asylum seekers who are situated in Stuttgart in the Baden-Württemberg region, who arrived in Germany after 2011/2012 and hold the protection or refugee status (refugees) in addition to those who are still waiting for obtaining refugee status (asylum seekers). Baden-Württemberg has been one of the regions that received large number of applications from asylum seekers in 2016 and 2017 with a Syrian majority (BAMF 2018). The study targeted both women and men aged between 18 and 64 years old. The interviews, all in Arabic language, covered the changes in their food-related habits before, during, and after migrating to Germany, and their perception of the new food environment. This case study provides insight to participants' perception of the changes in their food-related habits, the challenges they face in the new food environment, and the role of technology in enhancing their food and nutrition security.

8.4.1   Food During Wartime and Migration Journey

In wartime, food is sometimes used as a weapon to weaken a population by restricting its accessibility (Grinspan 2014). Although this is considered a war crime, it is still used nowadays. For instance, during the war in Syria, several areas were put under siege, in an attempt to suppress the population. The war in Syria has a disruptive impact on the local food systems, people face higher food prices, and reduced availability of several food items, specifically fresh fruits, and vegetables produce (Carnegie Endowment for International Peace 2015). The reduced availability and accessibility of food have had a dramatic impact on people's food and nutrition security. According to interview partners, being under tightening siege in the eastern Ghouta (the countryside and suburban area in southwestern Syria that surrounds the city of Damascus), pushed people to the verge of famine, where they were forced to *"eat everything bad and inedible"*, even including fodder and grass. Besides, some respondents horrifically witnessed some of their acquaintances starving to death as a result of severe malnutrition. Cutting supplies and shortages have led to skyrocketing prices, which pose extreme psychological stress for participants on how to feed their children, as illustrated in the following quote:

> When your child asks you for a Labneh[2] sandwich, and you say that we don't have bread, then (s)he will reply by asking for a Zaatar sandwich. The young child does not understand.

The struggle for food does not end by successfully fleeing the country. During the long and perilous migration route from Syria arriving in Germany, passing by several countries most likely Turkey, then by boat to Greece, passing by Macedonia, Serbia, Hungary, and Austria, refugees encountered reduced availability of food, irregular access to food, and/or low-quality diet. At that time, food was not a priority according to participants. They needed the minimum to have the force to continue their journey. However, they reported that in some cases food was thrown at them in a way that hurt their dignity. As one of the participants described:

> They were throwing food at us, I mean, I felt like we were animals, and they were throwing food at them.

The word "dignity" is hardly mentioned in the rhetoric around food in case of forced migration. Therefore, few participants found themselves in a position of trading off between being humiliated and feeding their children, as expressed:

> I have a self-pride when they bring food, I couldn't scramble to take it. If someone gives me food, then I will eat. I felt that my self-pride doesn't allow me to scramble to get food. At the same time, if I didn't do so, my children will remain without food. It's one of two things, either you humiliate yourself, or your children won't eat.

An important notion here to remind us that efforts to tackle food insecurity must center on preserving human dignity and ensuring an effective enforcement of the right to food approach, and not only on the traditional view on food security as an issue of food availability, and later on, of food accessibility (Koc 2013).

8.4.2 Adaptation to the New Food Environment

When moving to developed countries migrants might face multiple food-related challenges, such as forced changes in dietary patterns and little knowledge on how to use available resources. This new reality could have a dramatic impact on their food security situation. According to participants, the first period after arriving in Germany has daunting difficulties. Everything is entirely new and obscure, even the most rudimentary actions, such as purchasing food. Fortunately, this isolation between participants and the new food environment is slowly being alleviated. They gradually acquire the tools and skills that enable them to discover their surrounding environment. With many trials and tribulations, they acquire information about the existing food items. First, they look for the same kind of food that they used to have

---

[2] Labneh is a yoghurt-like Arabian sour milk.

back in their home country. Then, they start to make a comparison between what is available in the German food environment and what they used to have back in Syria. According to participants, the way of purchasing food has changed. Both men and women share shopping, but in case the supermarkets are far from their living place, then men are responsible for it. All of the young and unmarried participants who used to live with their family in Syria started to practice shopping after arriving in Germany. Moreover, while food shopping is typically done on a daily basis in Syria, with visits to multiple specialty shops, in Germany shopping is done less frequently with larger quantities of specific food items, such as flatbread, and halal meat. This is mainly because such items can only be purchased from oriental shops, which are few and far away from their living places. As an adaptation to the new living conditions, mobile food vehicles run by refugees started to circulate and bring several non-perishable food items to refugees' resettlement centers, allowing camp residents to save time and effort. One of the challenges refugees face is language barriers as recalled by several participants which affect their ability to make autonomous food choices. In some cases, participants reported their intention to purchase a specific food item, but having ended up buying something else as a result of their language deficiency. Other times, they are forced to buy some food items from ethnic shops with higher prices, either because they don't know what the items are called in German language, or because they cannot read the label to check for proscribed ingredients. Since the majority of Syrian are Muslims, eating according to religious practices is considered essential, although Muslims vary in the extent to which a halal diet is followed. As known, pork is totally avoided, and observant Muslims usually consume halal meat and do not drink alcohol. However, the new food environment for some participants is flooded with non-halal products, and therefore few participants are skeptical about consuming several food products, as explained in the following quote:

> Since I arrived here so far, I did not eat chips, did not eat chocolate, did not eat Mortadella. There are many items that I did not eat. I mean I love them so much [...] but I stopped eating them as I am afraid that they might contain pig [derivatives, or] Gelatin.

Another important aspect that Syrian refugees always use to evaluate food is taste. According to (the perception of) participants the German produce is devoid of taste and full of hormones and synthetic products. The majority of them describe the German produce as having "*only shape, no taste*" properties. However, the big size of vegetables (fruits are less prone to this) is what makes the interviewees think that agricultural produce is full of hormones and chemicals. This, in turns, leads to a feeling of mistrust of the healthfulness of German produce. To go further, participants asserted that when they prepare their typical food, the same way they used to do in Syria, using the same ingredients, the taste is not rich as it should be:

> Here we use the same method of preparation and the same ingredients but the taste is different.

The bigger size of vegetables was the result of multiple years of breeding "where a premium has been placed on heft" (Carolan 2017, p. 34). Since growers are paid for size and not for flavors, they started to grow bigger vegetables, which have fewer flavors and minerals, which is not the case back in Syria. Fresh fruits and some vegetables are seasonably available. Therefore, people used to purchase large quantities during the season and to preserve them for the unseasonal (winter) period, this process called "Elmouna". This has changed after arriving in Germany since these vegetables usually are not available fresh. In general, the German food environment with the existence of oriental groceries is abundant – "almost everything is available" – and the food is somehow easy to access. Nevertheless, experiencing financial stress was shared by participants who have large families and the ones who sent remittance to their family back in Syria, which has arguably affected their dietary patterns and family dynamics around food. For instance, the consumption of vegetables has reported to be less comparing to Syria as vegetable prices are much higher in Germany. On the contrary, meat prices become more affordable for them; therefore, their diet has changed, which has negatively affected their health, as explained by the following quote:

> In Germany, I get tired a lot when I walk or run a little bit. This is because [our diet has changed]. The nature of our food in Syria depends on vegetables. Vegetable prices are low in the season there. However, here vegetables are expensive. One would like to cook zucchini, and it is expensive, for example, a Stuffed zucchini costs 50 euros. Therefore, we rely mainly on rice and chicken.

Syrian migrants are like other migrants; the acculturation process that they go through has a substantial impact on their food security situation on multiple pathways. Purchasing food items from ethnic stores with higher prices, as a result of language deficiency, has a negative impact on home budget management. Moreover, paying more for food and sending remittance to family back in Syria, besides smoking for Syrian men who live alone in Germany, lead to the early completion of the income. In addition, language barriers and not having access to information about the available food items and their qualities undermine Syrians' abilities to make informed food choices which meet their values. This jeopardizes their food security situation by having reduced access to culturally accepted food that matches their preferences.

8.4.3   The Role of Technology in the Process of Adaptation to the New Food Environment

Food technology as we mentioned it so far does not play a significant role yet in the life of migrants, and the understanding of food technology has to undergo a semantic operation as to be enlarged to include technology food-related. Interestingly, food-related technology has a vital role in keeping migrants in touch with their traditional food products. Indeed, food processing and trading made a wide range of food items available through the migrant-run ethnic food retailers. These ethnic grocery stores support migrants to construct their new food environment actively and not only being passive by depending merely on the host countries' predominant food environments (Khojasteh and Raja 2017; Schiller and Caglar 2010). Making available a wide range of food items helps facilitating the process of preserving traditional food diet, which was crucial for our participants. Even in exile, food for Syrians still preserved its prestigious value, and they still talked about it with the same pride and obsession with 'Terroir" as the French do (Steavenson 2016). Therefore, the majority of participants still keep their food-related habits and have asserted their attachment to traditions and customs:

> [My future consumption] will not change over time, it will stay as it is. Because we have customs and traditions and there are things we cannot dispense with like Halal!

Furthermore, to adapt to the new lifestyle and to preserve the traditional food-related knowledge and heritage, new food-related social media and mobile apps have been created. New channels on YouTube and Facebook groups and pages have been developed by migrants from the same ethnic groups to inform the followers about the traditional food and the new food environment and to ease their life in the host country. These initiatives have been invented with multiple objectives: (1) Educational: to teach cooking skills for those who do not know how to cook. (2) Documenting: to document recipes and stories to typical dishes (e.g., Sham Alaseel YouTube channel,[3] for the first and second objectives). (3) Explorative: to inform followers about the existing products in the host country retailers and the places where they can find Arabian and Middle Eastern Cuisine (e.g., Pinda, mobile-app[4]). (4) Alerting: to inform followers about the existence of proscribed ingredients in various food items to avoid them (e.g., "Was ist das & Wofür?", Facebook group[5]).

Interestingly, all these initiatives have been created mainly by refugees in 2016 when the borders were opened, and a large number of them came to Europe. The difficulties they have encountered have pushed them to come up with solutions. These solutions were only possible by making use of modern technologies. Therefore, these initiatives could be seen as participatory and at the same time empowering for this group of migrants. Spreading food-related information and knowledge through the internet using the Arabic language has helped many participants to preserve their culinary habits. By using these food-related techs, they were able to prepare all the foods that they cannot find in Germany, such as yogurt, curd, labneh, makdous (tiny, tangy eggplants stuffed with walnuts, red pepper, garlic, olive oil, and salt), and pastries like Halawet el-jibn (Semolina and cheese dough, filled with cream), knafeh, etc. As one participant explains:

---

[3] Sham Alaseel channel created by a Syrian refugee in Sweden, where he presents the way of preparing traditional Syrian dishes and their stories.

[4] Pinda mobile-app: is a Syrian/German platform provides the locations of near oriental food places, such as restaurants, supermarkets, and oriental cuisine cookers.

[5] Was ist das & Wofür? Facebook women's group created by Syrians with different specialists to examine the list of ingredients of different products.

In Syria, we did not make pastries. Everything was available. Here [in Germany], those foods are not available. Therefore, I learned how to make all the sweets, such as Halawet El Jibn. Necessity is the mother of invention.

Additionally, by using these food-related techs, they were able to overcome language barriers and food illiteracy. Simply, by posting a photo of a product and its list of ingredients, they were able to know if the product is matching their values or not (in terms of proscribed ingredients). Therefore, by integrating these food-related technologies in their everyday life, they somehow become more familiar with the new food environment and more connected to it.

What is so special about these food-related technologies, is that they have been created out of need. This might be not so common in the food technology sector, where an entirely different top-down approach is followed by big companies behind development or profit-oriented ones. Therefore, these technologies have the potentials to solve broader problems faced by migrants, such as time and distance problem through applications that facilitate delivery services or connecting several people who can order together or teaching them how to make their diet healthier through social media or apps, and so forth.

The food technology played another role in the life of migrants, but this time was negative. Young participants who live alone in Germany, mainly men, because of the lacking of cooking skills, started to rely on convenience food for eating which is abundantly available in the German food environment, such as seasoned chicken, pasta, and pizza. This diet transition has led to serious health problems for some of them, as explained by one participant:

In Germany, I only look for quick things, because I cannot cook every day, only at public holidays. And on regular days, I look for things that are quick, such as pizza […]. But frankly, I suffer from this subject, and I was sick because of it. Fast food does not contain many vegetables, and this leads to indigestion. I consult a doctor about this subject and conducted a small operation. Therefore, the food here affects me a lot, and I always try, after what I have experienced, to eat a lot of vegetables.

Technology for Syrian migrants is not understood as having the latest high tech or consuming minimally processed food, but rather is used as a way of preserving their traditional food knowledge and keeping their traditional diet as a way of creating a sense of place and protecting their ethnic identity. The food industry still has a crucial role in making better options available for migrants, who are economically disadvantaged, to enable them to have better health and to express their identities.

8.5 Conclusion

The current global food system is constructed on the principles of productivity, efficiency, and profitability, and thus overlooks other features, such as social justice, sustainability, and sovereignty. This structure of the modern food system has had negative impacts not only on the environment but also on the health of the population, especially the low-income and disadvantaged strata of the society. Instead of addressing food security, this capitalist economy has been a source of food insecurity (Albritton 2009). Nevertheless, the discourse of food insecurity was effective for legitimizing their proliferation. The food industry, as one of the main actors of this system, focused on producing enough calories in highly efficient ways regardless of the type of calories produced or the way of producing them to attain massive profits. This has led to the spread of the processed food industry and culture all around the globe. Consequently, a wide range of malnutrition and non-communicable diseases, such as obesity, diabetes, and cardiovascular diseases, have started to be a new norm among particularly the low-income and disadvantaged population. The development of food technology could play a positive role in reversing the adverse effects that the modern system reproduced, but this can only happen if the industry does not merely focus on food availability or accessibility, but rather on enabling the social control over the food system and reducing social inequalities between affluent and disadvantaged populations including migrants. It is vital for the industry to keep food prices cheap enough to stay in the reach of the economically disadvantaged people, regardless of the expected hindrances resulted from the of Engel's law disruption. Furthermore, the industry should commit to producing healthier food instead of nutritiously shallow energy-dense ones, and most importantly to make these foods accessible to the whole strata of the society, and not just the affluent. Enhancing the food and nutrition security of disadvantaged and migrants' communities can be done by responding to their needs and facilitating their social inclusion and thus assisting them through their transitional process. According to what we have seen in the case study of Syrians refugees in Germany, food is a symbolic resource for migrants that they used to maintain a sense of continuity beyond changes. Hence, maintain traditional diet and seeking to revive the sensory experiences help migrants to make sense of the place and create like 'home' environment, which will eventually enhance their well-being and their integrational level. To make this possible, the industry could collaborate with experts and activists to develop processed food which meets migrants needs and sensory expectations in places where global distribution might struggle. Furthermore, the food industry should use technologies to make their promises of a better life for everyone come true. Making food technology available and accessible at the household level is essential, but not without using food-related technologies to educate users and to facilitate the process of food choices. To address the problem of food insecurity, it is crucial for the industry to create an enabling environment and to empower the whole strata of society by giving them tools, skills, and resources to navigate their food environment and to make informed, personalized and socially-accepted food choices that meet their values and goals, and respect their dignity. The advancement of food technology and food digitalization is unstoppable, therefore is inevitable to orient this evolution to achieve food security and social inequalities, which can be done by focusing on the food-related technologies that meet the needs of all the strata of society and not overtaken the impoverished ones. By listening to their needs, the industry will be able to create sustainable and inclusive food system and participate in creating a healthier individual and a healthier planet.

References

Albritton, R. (2009). *Let them eat junk: How capitalism creates hunger and obesity*. London: Pluto Press. Resource document. https://www.jstor.org/stable/j.ctt183pbv8.

BAMF. (2018). Asylgeschäftsbericht 12/2017. Resource document. https://www.bamf.de/SharedDocs/Anlagen/DE/Downloads/Infothek/Statistik/Asyl/201712-statistik-anlage-asyl-geschaeftsbericht.html;jsessionid=2FFEC95302F681CD3C054EB165B507A3.2_cid294?nn=1367522. Accessed 11 Oct 2018.

Barry, D. T., & Garner, D. M. (2001). Eating concerns in east Asian immigrants: Relationships between acculturation, self-construal, ethnic identity, gender, psychological functioning and eating concerns. *Eating and Weight Disorders – Studies on Anorexia, Bulimia Obesity, 6*(2), 90–98. https://doi.org/10.1007/BF03339757.

Bey, M. (2016). Between geopolitics and technology. Retrieved August 12, 2018, from https://worldview.stratfor.com/article/between-geopolitics-and-technology.

Braverman, H. (1998). *Labor and monopoly capital: The degradation of work in the twentieth century*. New York: NYU Press.

Brown, L. R. (2012). *Full planet, empty plates: The new geopolitics of food scarcity*. New York/London: W.W. Norton.

Burns, C. (2004). Effect of migration on food habits of Somali women living as refugees in Australia. *Ecology of Food and Nutrition, 43*(3), 213–229. https://doi.org/10.1080/03670240490447541.

Butz, P., & Tauscher, B. (2002). Emerging technologies: Chemical aspects. *Food Research International, 35*(2), 279–284. https://doi.org/10.1016/S0963-9969(01)00197-1.

Carnegie Endowment for International Peace. (2015, June 4). Food insecurity in war-torn Syria: From Decades of Self-Sufficiency to Food Dependence. Resource document. https://carnegieendowment.org/2015/06/04/food-insecurity-in-war-torn-syria-from-decades-of-self-sufficiency-to-food-dependence-pub-60320. Accessed 2 Oct 2018.

Carolan, M. S. (2013). *Reclaiming food security*. Abingdon: Routledge.

Carolan, M. S. (2017). *No one eats alone: Food as a social enterprise*. Washington, DC: Island Press/Center for Resource Economics. Resource document. www.springer.com/de/book/9781610918060.

Carroll, J. K., Moorhead, A., Bond, R., LeBlanc, W. G., Petrella, R. J., & Fiscella, K. (2017). Who uses mobile phone health apps and does use matter? A secondary data analytics approach. *Journal of Medical Internet Research, 19*(4), e125. https://doi.org/10.2196/jmir.5604.

Carswell, K., Blackburn, P., & Barker, C. (2011). The relationship between trauma, post-migration problems and the psychological well-being of refugees and asylum seekers. *International Journal of Social Psychiatry, 57*(2), 107–119. https://doi.org/10.1177/0020764009105699.

Chadwick, J. (2017). Here's how 3D food printers are changing what we eat. Resource document. https://www.techrepublic.com/article/heres-how-3d-food-printers-are-changing-the-way-we-cook/. Accessed 10 Aug 2018.

Coughlin, S. S., Whitehead, M., Sheats, J. Q., Mastromonico, J., Hardy, D., & Smith, S. A. (2015). Smartphone applications for promoting healthy diet and nutrition: A literature review. *Jacobs Journal of Food and Nutrition, 2*(3), 021.

Counihan, C., & Esterik, P. V. (2013). *Food and culture: A reader*. New York/Abingdon, Oxon: Routledge.

Depa, J., Gyngell, F., Müller, A., Eleraky, L., Hilzendegen, C., & Stroebele-Benschop, N. (2018). Prevalence of food insecurity among food bank users in Germany and its association with population characteristics. *Preventive Medicine Reports, 9*, 96–101. https://doi.org/10.1016/j.pmedr.2018.01.005.

Dharod, J. M., Croom, J., Sady, C. G., & Morrell, D. (2011). Dietary intake, food security, and acculturation among Somali refugees in the United States: Results of a pilot study. *Journal of Immigrant & Refugee Studies, 9*(1), 82–97.

Dinour, L. M., Bergen, D., & Yeh, M.-C. (2007). The food insecurity–obesity paradox: A review of the literature and the role food stamps may play. *Journal of the American Dietetic Association, 107*(11), 1952–1961. https://doi.org/10.1016/j.jada.2007.08.006.

Dixon, J. (2009). From the imperial to the empty calorie: How nutrition relations underpin food regime transitions. *Agriculture and Human Values, 26*(4), 321. https://doi.org/10.1007/s10460-009-9217-6.

Drewnowski, A., & Specter, S. E. (2004). Poverty and obesity: The role of energy density and energy costs. *The American Journal of Clinical Nutrition, 79*(1), 6–16. https://doi.org/10.1093/ajcn/79.1.6.

Engel, E. (1857). Die Produktions- und Consumtionsverhaltnisse des Königreichs Sachsen. *Zeitschrift Des Statistischen Bureaus Des Königlich Sächsischen Ministeriums Des Innern, 8* and *9* (Reprinted in Engel (1895). Appendix I, pp. 1-54.).

Eurostat. (2018). Asylum statistics – Statistics explained. Resource document. https://ec.europa.eu/eurostat/statistics-explained/index.php/Asylum_statistics#undefined. Accessed 2 Oct 2018.

FAO. (2017). SOFI 2017 – The State of Food Security and Nutrition in the World. Resource document. http://www.fao.org/state-of-food-security-nutrition/en/. Accessed 10 Aug 2018.

Fennelly, K. (2007). The "healthy migrant" effect. *Minnesota Medicine, 90*(3), 51–53.

Floros, J. D., Newsome, R., Fisher, W., Barbosa-Cánovas, G. V., Chen, H., Dunne, C. P., et al. (2010). Feeding the world today and tomorrow: The importance of food science and technology. *Comprehensive Reviews in Food Science and Food Safety, 9*(5), 572–599. https://doi.org/10.1111/j.1541-4337.2010.00127.x.

FMI, & Salmon, K. (2017). Future of food: New realities for the industry. Accenture Strategy. Resource document. https://www.fmi.org/forms/store/ProductFormPublic/future-of-food-new-realities-for-the-industry.

Fowokan, A. O., Black, J. L., Holmes, E., Seto, D., & Lear, S. A. (2018). Examining risk factors for cardiovascular disease among food bank members in Vancouver. *Preventive Medicine Reports, 10*, 359–362. https://doi.org/10.1016/j.pmedr.2018.04.015.

Franco, R. Z., Fallaize, R., Lovegrove, J. A., & Hwang, F. (2016). Popular nutrition-related Mobile apps: A feature assessment. *JMIR MHealth and UHealth, 4*(3), e85. https://doi.org/10.2196/mhealth.5846.

Geissler, C., & Powers, H. (2010). *Human nutrition – E-book*. Edinburgh: Elsevier Health Sciences.

Grinspan, J. (2014). How coffee fueled the civil war. Resource document. https://opinionator.blogs.nytimes.com/2014/07/09/how-coffee-fueled-the-civil-war/. Accessed 2 Oct 2018.

Hadley, C., Zodhiates, A., & Sellen, D. W. (2007). Acculturation, economics and food insecurity among refugees resettled in the USA: A case study of West African refugees. *Public Health Nutrition, 10*(4), 405–412. https://doi.org/10.1017/S1368980007222943.

Hadley, C., Patil, C. L., & Nahayo, D. (2010). Difficulty in the food environment and the experience of food insecurity among refugees resettled in the United States. *Ecology of Food and Nutrition, 49*(5), 390–407. https://doi.org/10.1080/03670244.2010.507440.

Hassan, D. A. E., & Hekmat, S. (2012). Dietary acculturation of Arab immigrants: In the Greater Toronto area. *Canadian Journal of Dietetic Practice and Research, 73*(3), 143–146. https://doi.org/10.3148/73.3.2012.143.

Hersey, J., Anliker, J., Miller, C., Mullis, R. M., Daugherty, S., Das, S., et al. (2001). Food shopping practices are associated with dietary quality in low-income households. *Journal of Nutrition Education, 33*(Suppl 1), S16–S26.

HNO. (2018). Humanitarian needs overview 2018 – Syria Crisis. Resource document. http://hno-syria.org/. Accessed 23 April 2018.

Holmboe-Ottesen, G., & Wandel, M. (2012). Changes in dietary habits after migration and consequences for health: A focus on South Asians in Europe. *Food & Nutrition Research*, (*56*). https://doi.org/10.3402/fnr.v56i0.18891.

IMO. (2017). *World migration report 2018*. Switzerland. Resource document. https://www.iom.int/wmr/world-migration-report-2018.

Jackson, T. (2009). *Prosperity without growth: Economics for a finite planet*. London/New York: Earthscan.

Jaffe, J., & Gertler, M. (2006). Victual vicissitudes: Consumer deskilling and the (gendered) transformation of food systems. *Agriculture and Human Values, 23*(2), 143–162. https://doi.org/10.1007/s10460-005-6098-1.

Jetter, K. M., & Cassady, D. L. (2006). The availability and cost of healthier food alternatives. *American Journal of Preventive Medicine, 30*(1), 38–44. https://doi.org/10.1016/j.amepre.2005.08.039.

Khojasteh, M., & Raja, S. (2017). Agents of change: How immigrant-run ethnic food retailers improve food environments. *Journal of Hunger & Environmental Nutrition, 12*(3), 299–327. https://doi.org/10.1080/19320248.2015.1112759.

Koc, M. (2013). Food security as a neo-liberal discourse. In *Accumulations, crises, struggles: Capital and labour in contemporary capitalism* (pp. 245–265). Berlin: LIT Verlag.

Koc, M., & Welsh, J. (2002). Food, foodways and immigrant experience. Presented at the Canadian Ethnic Studies Association Conference, Halifax: Centre for Studies in Food Security.

Kwik, J. (2008). Traditional food knowledge: A case study of an immigrant Canadian "foodscape". *Environments: A Journal of Interdisciplinary Studies, 36*(1), 6.

Lawrence, G., & McMichael, P. (2012). The question of food security. *International Journal of Sociology of Agriculture and Food, 19*(2), 135–142.

Lawson, M., Chan, M.-K., Rhodes, F., Parvez Butt, A., Marriott, A., Ehmke, E., et al. (2019). Public good or private wealth? (p. 106). Oxfam. https://doi.org/10.21201/2019.3651

Lesser, I. A., Gasevic, D., & Lear, S. A. (2014). The association between acculturation and dietary patterns of south Asian immigrants. *PLoS One, 9*(2), e88495. https://doi.org/10.1371/journal.pone.0088495.

Mackert, M., Mabry-Flynn, A., Champlin, S., Donovan, E. E., & Pounders, K. (2016). Health literacy and health information technology adoption: The potential for a new digital divide. *Journal of Medical Internet Research, 18*(10). https://doi.org/10.2196/jmir.6349.

MacroGeo, & Barilla Center for Food & Nutrition Foundation (BCFN). (2017). *Food and migration – understanding the geopolitical Nexus in the Mediterranean Area*. Parma -Italy. Resource document. https://www.barillacfn.com/m/publications/food-and-migration.pdf.

Mudry, J. (2006). Quantifying an American eater. *Food, Culture & Society, 9*(1), 49–67. https://doi.org/10.2752/155280106778055172.

Nair, T. (2016). 3-D printing for food security: Providing the future nutritious meal. *RSIS Commentaries, 273*(16). Resource document. https://dr.ntu.edu.sg/handle/10220/41717

Oussedik, S. (2012). Food and cuisine: Part of the migration process. *Institut Europeu de la Mediterrània*. Resource document. http://www.iemed.org/observatori/arees-danalisi/arxius-adjunts/qm-17-originals/qm17_Oussedik.pdf.

Patil, C. L., McGown, M., Nahayo, P. D., & Hadley, C. (2010). Forced migration: Complexities in food and health for refugees resettled in the United States. *NAPA Bulletin, 34*(1), 141–160. https://doi.org/10.1111/j.1556-4797.2010.01056.x.

Pérez-Escamilla, R., Ferris, A. M., Drake, L., Haldeman, L., Peranick, J., Campbell, M., et al. (2000). Food stamps are associated with food security and dietary intake of inner-city preschoolers from Hartford, Connecticut. *The Journal of Nutrition, 130*(11), 2711–2717. https://doi.org/10.1093/jn/130.11.2711.

Rosenblum, M. R., & Tichenor, D. J. (2012). *The Oxford handbook of the politics of international migration*. Oxford/New York: OUP.

Rosin, C., Stock, P., & Campbell, H. (2013). *Food systems failure: The global food crisis and the future of agriculture*. London/New York: Routledge.

Schiller, N. G., & Caglar, A. (Eds.). (2010). *Locating migration: Rescaling cities and migrants*. Ithaca: Cornell University Press.

Simmet, A., Tinnemann, P., & Stroebele-Benschop, N. (2018). The German food bank system and its users-a cross-sectional study. *International Journal of Environmental Research and Public Health, 15*(7). https://doi.org/10.3390/ijerph15071485.

Steavenson, W. (2016). Remembrance of tastes past: Syria's disappearing food culture | Wendell
    Steavenson. The Guardian. Resource document. http://www.theguardian.com/world/2016/
    dec/07/syria-refugees-disappearing-food-culture-kibbeh.

Swinburn, B., & Egger, G. (2002). Preventive strategies against weight gain and obesity. *Obesity
    Reviews: An Official Journal of the International Association for the Study of Obesity, 3*(4),
    289–301.

UNHCR. (2018). Forced displacement at record 68.5 million. Resource document. http://www.
    unhcr.org/news/stories/2018/6/5b222c494/forced-displacement-record-685-million.html.
    Accessed 10 Aug 2018.

West, J. H., Belvedere, L. M., Andreasen, R., Frandsen, C., Hall, P. C., & Crookston, B. T. (2017).
    Controlling your "app"etite: How diet and nutrition-related Mobile apps lead to behavior
    change. *JMIR MHealth and UHealth, 5*(7), e95. https://doi.org/10.2196/mhealth.7410.

WHO. (2018). Malnutrition. Resource document. http://www.who.int/news-room/fact-sheets/
    detail/malnutrition. Accessed 10 August 2018.

Wrangham, R. (2009). *Catching fire: How cooking made us human*. New York: Basic Books.

Chapter 9
Corporate Responsibility in a Transitioning Food Environment: Truth-Seeking and Truth-Telling

Louise Manning

9.1 Introduction

This book chapter considers the role of corporate responsibility in a transitioning food environment. The chapter starts by contextualizing the transitioning food environment, before critiquing the role of corporate responsibility and the delivery of brand value in a food market with a growing dynamic of truth-seeking and what this means for corporate truth-telling. The drivers of transition are economic, socio-political, technological, and environmental especially as global resources come under more pressure. Terms such as truth values, social trust, corporate honesty, truth seeking and truth telling are explored. Transitioning from the prevailing food regime to another will be driven by multiple factors primarily consumers seeking truth and transparency with the foods they purchase and consume and ecopreneurial niches affecting existing regime path dependency and an unwillingness to change. This evolving socio-technical symbiosis will be required to meet the pressures and socio-economic drivers that influence the food supply chain.

9.2 A Transitioning Food Environment

In order to meet global population needs in conjunction with ecological and social equity goals, a global scale societal transition is required especially in urban environments (Cohen and Ilieva 2015). *Nutrition transition* is often defined as the increasing share of animal protein in the diet (Popkin 2001), but can also describe the dietary shift towards a diet high in fat, refined foods and low in fibre (Popkin 2002) with the latter causing increased incidence of obesity, cardiovascular disease,

L. Manning (✉)
School of Agriculture, Food and Environment, Royal Agricultural University, Cirencester, UK
e-mail: louise.manning@rau.ac.uk

© Springer Nature Switzerland AG 2019
C. Piatti et al. (eds.), *Food Tech Transitions*,
https://doi.org/10.1007/978-3-030-21059-5_9

type 2 diabetes and cancer (Popkin 2007, 2008; Burggraf et al. 2015). By 2050, these predicted dietary trends, could be a major contributor to an estimated 80% increase in greenhouse gas emissions (GHGEs) from food production and land clearing (Dwivedi et al. 2017:844). However, human diets and associated food choices do not follow a linear trend and alter over time via a range of factors including evolving social norms (Alexander et al. 2016). Therefore, rather than operating in one plane and being driven by cause and effect dynamics such social norms change, adapt, transition and evolve at the system level. *Sustainability transitions* are social change processes that are non-linear through which a societal system is physically transformed (Avelino and Rotmans 2009). A sustainable transition towards less societally impactful diets that meet global population rise would promote economic return through production efficiency whilst minimizing environmental and social consequences (Ranganathan et al. 2016; Liao and Brown 2018).

Alternatively, sustainability transition has been described as the "radical transformation towards a sustainable society as a response to a number of persistent problems confronting contemporary modern societies" (Grin et al. 2010:1). Concern over depletion of natural resources and pollution has driven organizations, and by inference their supply base, and consumers to change the way they produce, purchase and consume food in order to deliver sustainable development (Jaca et al. 2018). This sustainable development can be enacted independently and in a mutually concerted way at the individual, household, community, national, regional and global scale. However, whilst technological improvements have driven efficiency reductions in emissions and improvements in the use of resources including water and energy, those benefits have largely been offset by increasing production and consumption volumes (Vergragt et al. 2016).

Transition governance can be based on limited normative target setting by single actors, but an alternative approach would be that the aims and objectives of human transition to sustainable diets is a negotiated space and defined, advocated and articulated by the actors involved (Rauschmayer et al. 2015). However, to drive population transition towards a sustainable diet some governments do set specific targets for national consumption, for example in Norway the government set a target that organic food would account for 15% of food consumption by 2020 (Vittersø and Tangeland 2015). Promoting more sustainable diets requires much more attention to cultural and social contexts such as taste, competencies and skills to prepare and cook food, and social relatedness and how they frame the foods we consume (Schösler and de Boer 2018). One social driver in some societies is to reduce GHGEs by eating less meat in the midst of a global transition towards more meat consumption. The term *flexitarian* has been used to describe individuals who are meat eaters, but choose to eat meat less than 4 days a week (Dagevos and Voordouw 2013) and this partial substitution by consumers suggests a trend towards more sustainable diets (Schösler and de Boer 2018).

Despite the term *transition* being widely used in the literature its meaning is unclear (Silva and Stocker 2018). Transition has been described as a period of unresolved experimentation and contestation (Levidow 2015) or a set of connected changes or developments concerning technology and materials, economy, political and institutional drivers, organizational, socio-cultural, ecology and belief systems, that coevolve or reinforce each other, but take place in several different areas, levels and time frames (Rotmans et al. 2001; Rotmans and Loorbach 2009; Markard et al. 2012). Transitions are thus co-created involving a broad range of actors (Markard et al. 2012) and they shift an existing regime from one particular socio-technical configuration towards another (Rauschmayer et al. 2015). Transitioning socio-technical food systems toward sustainability involves changing the materials and practices that produce and reframe them (Cohen and Ilieva 2015). This chapter seeks to contextualise the transitioning food environment, and then critiquing the role of corporate responsibility and the delivery of brand value in a food market within a growing dynamic of truth-seeking and what this means for corporate truth-telling.

9.3   Sociotechnical Systems

The term "socio-technical" was first used in the 1940s when considering the association between technical and social systems (Ghaffarian 2011). Socio-technical systems are the systems that involve a complex interaction between humans, machines and environmental aspects of the working system (Emery and Trist 1960, 1965). Whether the issue is meeting the United Nations (UN) Sustainable Development Goals, addressing the uptake of technology in the food supply chain, labour and human rights issues, food safety, nutrition and health, reducing food waste, or addressing corruption, crime and food fraud; this socio-technical interaction mediates all activity and responses by the food supply chain. Winter et al. (2014:251) propose that a key consideration with socio-technical systems is that "technologies themselves are not deterministic, but rather their impacts arise from complex interactions with industrial and organizational contexts". However, Winter et al. (2014) argue that organizations do not sit in isolation they interact with external influences (environment). The objective of socio-technical systems is the joint optimization of both the social and technical systems; the relationship between the two systems and the relationship with the external environment (Mumford 2006). Key characteristics of socio-technical systems are that they:

- Are interdependent:
- Have an internal environment that encompasses separate but interdependent technical and social subsystems where performance relies on the optimization of both the technical and social subsystems;
- Can adapt to and pursue goals in external environments often driven by supply chain actors and stakeholders;
- Have the characteristic of equi-finality that is the systems goals can be achieved by more than one means; and
- Have the limiting aspect that focusing on one system (e.g. technical) to the exclusion of the other (social) is likely to lead to degraded system performance and utility (Badham et al. 2001).

Socio-technical systems transition when multiple societal and technical factors co-evolve and shape each other in a non-linear approach (Silva and Stocker 2018). Therefore, socio-technical systems develop over many decades, and the alignment of their elements and external driving factors can lead to path dependence and resistance to change (Geels et al. 2017).

9.4   Regimes, Niches and Transition

Socio-technical transition results from the interaction between three system levels: the exogenous landscape, the prevailing regime (the mainstream), and the niche (Geels 2002). The *exogenous (external) landscape* of transition has already been described in this book chapter. It includes large scale socio-technical trends such as:

- Population growth, rapid urbanization, neoliberal globalization with more mobile capital and market deregulation (Smith et al. 2005);
- Dietary shift in some cultures towards more meat and dairy products, and in other cultures towards energy dense rather than nutrition rich foods, environmental and geopolitical crises (Kuokkanen et al. 2018); or
- Political ideologies, societal values, and economic patterns (Laakso and Lettenmeier 2016).

A *regime* can be described as a conglomerate of *structure* (institutional and physical setting), *culture* (prevailing perspective), and *practices* (rules, routines, and habits) (Rotmans and Loorbach 2009). Alternatively, a food regime is determined as a 'rule-governed structure of production and consumption of food on a world scale' (Friedmann 1993:30–31). Regimes exist across different empirical scales, with a high level of aggregation and are driven by rules and practices within a centralized system that is mediated or reinforced by consumer behaviour (Smith et al. 2005). The corporate food regime drives primary production to intensify production methods, and as a result "generating environmental harm, social inequalities and conflicts, especially in low-income countries" (Levidow 2015:78) thus the corporate-industrial and the corporate-environmental models have evolved as alternative (sub-set) regimes. Thus the overarching regime can be supported by "nested subordinate regimes" that underpin the status quo rather than being emerging niches where new technology and knowledge or new practices drive operational activities (Smith et al. 2005).

Put simply, niches are locations where it is possible to deviate from the rules in the existing regime (Geels 2004) or the locus for radical innovation (Geels 2010). A niche adopts new practices, technologies or ways of organizing activities and are places where mutual understanding among stakeholders can be built, aligned with collaborative goals and associated social and reflexive learning and unlock radical innovation (Meynard et al. 2017). Niches operate fully or at least partially outside the regime and offer an alternative socio-technical system often with a novel or innovative approach for example alternative food networks (AFNs) see Maye and

Kirwan (2010), technological food production innovations and novel food products (Kuokkanen et al. 2018). Niche innovations are emerging social or technical innovations that differ radically from the prevailing socio-technical system and regime, but are able to gain a foothold in particular applications, geographical areas, or markets or with the help of targeted policy support (Geels et al. 2017:465). Thus niches, such as food redistribution, are able to address a specific landscape pressure (food waste) better than the existing regime (Kuokkanen et al. 2018). Food redistribution AFNs can reconfigure, respatialise and resocialise the socio-technical systems of production, distribution and consumption of food (Jarosz 2008; Paül and McKenzie 2013). Thus, niches themselves can exert adaptive pressure on existing regimes (Smith et al. 2005). Regime change is a function of three processes:

- The first the impact of selection pressures on the regime;
- The second is the adaptive capacity of the regime and the coordination and allocation of resources available inside and outside the regime to adapt to these pressures by regime members, and
- Finally the impact of the regime change on the original selection pressures (Smith et al. 2005).

Adaptive capacity involves the use of resources to respond to selective pressures for change and as a result the degree of adaptive capacity required in a given situation is scaled to the selective pressures involved (Smith et al. 2005). Central to a regime is the focus of power on the status quo, political tactics and the role of incumbency (Smink et al. 2015), formation of coalitions, and the process of regime resistance (Geels 2004) i.e. slowing the pace of transition from a given situation (Johnstone and Newell 2018). Transitions to a new norm are difficult as "existing regimes [are] characterized by lock-in and path dependence, and oriented towards incremental innovation along predictable trajectories" (Geels 2010; Rauschmayer et al. 2015). Thus, the multi-level perspective (MLP) suggests that socio-technical system levels relate to heterogeneous power, coercive rule structure, pressure and structural dynamics (Kuokkanen et al. 2018) so power and agency are important drivers in regime transition and power facilitates or alternatively confines agency (Smith et al. 2005). Thus in the face of dominant actor strategies and their associated networks, knowledge and information asymmetry and locked in practices such path dependency does not encourage radical innovation and the designing of new ways of doing can be prevented by the prevailing regime (Geels 2002; Meynard et al. 2017).

System-level change requires the coordination of many actors and resources (Smith et al. 2005) and a co-evolution of new ideas and innovation. Indeed, radical innovations emerge when actors reach consensus, alignment and develop workable solutions (Geels 2010; Rauschmayer et al. 2015). Geels et al. (2017:466) assert:

> Struggles between niche innovations and existing regimes typically play out on multiple dimensions, including: economic competition between old and new technologies; business struggles between new entrants and incumbents; political struggles over adjustments in regulations, standards, subsidies, and taxes; and discursive struggles over problem framings and social acceptance.

Regime destabilization is "the moment when a 'window of opportunities' occurs for new niches to break through or when the selection environment is open for reformulation" (Kuokkanen et al. 2018:1515). Smith et al. (2005) define four kinds of socio-technical system response building on the original typology of Berkhout et al. (2004):

(a) **Endogenous renewal** (coordinated systems based response by the regime to selection pressures that involves clear internal articulation and adaption that leads to incremental change that follows a pre-determined path);
(b) **Re-orientation of trajectories** (uncoordinated response often to supply chain shock that may be endogenous or exogenous to the regime as a selection pressure that drives transition involving internal adaption that may be poorly articulated);
(c) **Emergent transformation** (uncoordinated response as a result of uncoordinated selection pressures for change and responses based on resources and capability outside the regime i.e. a response that involves external adaption);
(d) **Purposive transitions** (coordinated response primarily negotiated between actors outside the regime both in terms of articulating selection pressures for change and in providing resources, capabilities and networks i.e. a response that involves external adaption).

However how do these responses occur, what is the driver to driving transition? How does transition deliver to the actors involved? Socio-technical system transition is effective when it informs the prevailing regime to shift from one particular socio- technical configuration to another (Rauschmayer et al. 2015). This shift can occur via three independent or interlinked dynamics:

(a) Top down, when the exogenous landscape put adaptive pressure on the regime;
(b) Bottom up, when niches scale up and replicate more widely their innovations and gain influence; and,
(c) When regime level processes drive adoption of innovations from the niche level into the regime (Rotmans and Loorbach 2010; Rauschmayer et al. 2015).

Language, as a form of social power has an especially crucial role in shaping behaviour, processes and realignment of power relations and actors will use various discourses during the transition process to suit their needs (Lawhon and Murphy 2012). The first element of language is determining the innate characteristics of truth.

9.5   Truth Values

When determining what is or is not true, we have to consider whether truth is an absolute characteristic or whether truth is a relative and socially constructed term. Contemporary literature speaks to consideration of the lived language functioning pragmatically compared to the formalised language (Nørreklit et al. 2018) and there being a truth gap between proactive and pragmatic truth, or indeed perceptions of there being multiple truths depending on the viewpoint of the observer (Krzyzaniak 2018). Indeed, Krzyzaniak (2018) differentiates between an objective "one truth" mindset and a relativist "multiple truth" mindset that can be evidenced in the food supply chain. Therefore, truth is not only determined by what is tested or defined as being fact (objectively), but also by what people "mean" by the term, truth (Snellman 1911) i.e. truth as described by Krzyzaniak (2018) can be framed from a subjective, relativist viewpoint. Truth is linked to factors such as knowledge and reality, uncertainty and temporality (Andrikopoulos 2015). Indeed, Deaver (1990) describes *facts* as documentable elements of truth and *information* as a collection of related factors organized in a specific structure. Information asymmetry can occur between individuals or organizations, influencing concepts of truth and trust in the information conveyed, and whether it is acceptable to deceive or indeed fail to inform (Rosenbaum et al. 2014). Andrikopoulos (2015) highlights firstly notions of *correspondence* i.e. truth is the interaction between the believer and what is believed; and secondly notions of *coherence* whereby truth can vary from person to person and across time (see Davidson 2001), i.e. temporality affects what could be seen as "the truth" sometimes reflected on as being hindsight (Krzyzaniak 2018).

Whilst failure to disclose given information may be seen by some as being less concerning than actively and intentionally misleading others i.e. "not telling the truth" (Green and Kugler 2012) this is open to interpretation. Therefore, perceptions of corporate responsibility, blameworthiness and accountability are important, particularly when consumer confidence has been lost or damaged (Regan et al. 2015). *Objective truth* relates to facts whereas constructivist or *pragmatic truth* suggests that truth is co-created through social consensus i.e. that what is considered as truth is cognitively determined by mutual consent amongst individuals (Andrikopoulos 2015) and has been considered when critiquing corporate financial reporting (Mitchell et al. 2017). Fisher (2013) argued that truth is more complex than simply being the opposite of lying and sought to create a typology for truth (Table 9.1).

Table 9.1 Types of truth

Type of truth	Definition
Objectivist, verifiable approach to truth	A belief is true if and only if it corresponds to a fact
Positivist approach to truth:	Something is true because it is not false
Constructivist, interpretivist, pragmatic approach to truth	Truth is socially constructed, open to interpretation, causal explanation and dependent on individual perception
Selective approach to truth	Truth, but not the whole truth i.e. truth is constructed in terms of a narrative and is selective in terms of what is included and excluded in the definition of truth

Adapted from Berger and Luckmann (1991), Blumer (1986), Glanzberg (2014), Andrikopoulos (2015), and Fisher (2013)

A "half-truth" is "the communication of technically correct, truthful information that has been, or has the potential to be, undermined by the omission of key information" (Devin 2016:226) suggesting that truth can be partial or incomplete and yet still be deemed to be truthful. Deaver (1990) states too that there are various degrees of truth, half-truth and untruths (fiction) and proposes that truth is constructed on a continuum from:

(a) An intent to inform and be open, accurate and fully honest with no apparent bias,
(b) An intent to be honest but with selective use of information i.e. truth, but not the whole truth, often with the intent to persuade,
(c) The use of untruths, but with no intent to deceive, and
(d) A conscious intent to deceive.

Truth, however defined, affects trust and is now considered in the context of corporate behaviour, social trust and truth-telling.

9.6 Trust, Social Trust and Corporate Honesty

Trust is a key factor that mediates consumer perceptions of the environmental impact of production or social issues such as animal welfare, and the use of technology (Goddard et al. 2018). Further Krzyzaniak (2018) argues that trust can be seen by organisations as a construct that is a value or target. Trust can be defined as *general trust* i.e. relational, interactional-based trust between individual (micro-level) supply chain actors or groups of actors (macro-level). These actors include farmers, manufacturers and processors, pharmaceutical and agri-product companies, feed suppliers, and retailers or food service, advocacy groups (e.g. environmental groups or animal welfare groups) who act individually or cooperatively or in consort (Manning 2018a). Secondly, *individual trust* is formed at the individual business level between supply chain actors, or at the industry or institutional level with the government and/or experts and scientists (Hartmann et al. 2015; Charlebois et al. 2016; Goddard et al. 2018).

Social trust frames individual and collective actor behaviour in the supply chain as well as the informal governance of food policy in a society (Cao et al. 2016; Manning 2018a). Indeed, they argue that strong social trust dynamics create a trustworthy business environment so even if individuals are innately neither honest nor trustworthy they are likely to behave in a trustworthy way if they are under a social pressure to conform to such behaviour. Conversely if that social pressure does not exist such individually will behave negatively and a corrosive culture will occur in the organization and wider supply chain. Thus, social norms play a crucial role in the functioning of any socio-economic system, in particular the norms of trust and honesty and what is perceived as dishonesty (Galeotti et al. 2017). Cultural relativity influences what is defined as deception (Choi et al. 2011), with both objective and subjective norms framing deception with the latter having the stronger influence.

Institutionalization especially within transnational corporations has made supply chain activities less visible especially where regulation and market instruments operate at individual organization rather than whole supply chain level (Ali and Suleiman 2018), with third party certification often used as a proxy for trust (Manning 2018b). Trust reduces complexity in difficult and complex decision-making and is a cue for managing risk and uncertainty (Hartmann et al. 2015) i.e. trust is the stimulus through which 'social relations become productive' (Fisher 2013:15). Fisher in her work describes multiple attributes of trust including caring and concern, objectivity and fairness, openness and honesty, knowledge and expertise, information accuracy, credibility, shared goals, predictability (Peters et al. 1997; Kasperson and Golding 1992).

Honesty is not a fixed trait (Rosenbaum et al. 2014). Consumers value corporate honesty above corporate social responsibility behaviours (O'Connor and Meister 2008). For example, vague environmental claims have led to consumers questioning corporate honesty (Furlow 2010). Chance et al. (2015) suggest there are corporate trade-offs and it is better to consider a corporation's behaviour in terms of the business operating situational honesty i.e. that the organization will exhibit different approaches to honesty depending on the given scenario. Indeed, when corporate honesty is transparent to consumers they are more likely to purchase from the organization and this relationship is strong enough to lead to competitive advantage through increased turnover rather than higher prices (Pigors and Rockenbach 2016).

In the event of non-verifiable, and discretionary rather than mandatory corporate disclosure, credibility is an inherent problem (Dobler 2008). When information is non-verifiable, or the verification cost prohibitive, individuals can misreport and misrepresent the information and the intentions of the organization. Dobler (2008) described this as a "cheap talk" model that threatens perceptions of corporate credibility. Misrepresentation of information may be perceived as untruthful or lying if one individual intentionally communicates incorrect information to increase his or her benefit at the expense of others (Gneezy 2005; Schreck 2015). However, some actors may see misrepresentation as a moving feast and approach truthfulness as a subjective construct. Intention of malice is not an essential element of lying (Grover 1993). Factors that influence the propensity for lying are intra and inter-organizational competition and rivalry, asymmetry in information and resource allocation, monetary and/or status benefit for lying, and gender whereby men are influenced differently by competition and rivalry compared to women (Shreck 2015). However, there can be different kinds of lies and the corporate intention behind lying can vary. Lupoli et al. (2018:32) define *paternalistic lies* as "lies that are intended to benefit the target, but require the deceiver to make assumptions about targets' best interests" i.e. the perpetrator assumes it is in the best interest of the beneficiary to lie.

Paternalistic lies are a sub-set of *prosocial* lies used in the interest of the beneficiary, sometimes called "white lies". They are subjective and ubiquitous and reduce autonomy, and free will. Indeed, by intentionally deceiving others in an organization, certain individuals may believe they are acting prosocially (Grover 1993). Levine et al. (2018) in their study (n = 3883) found that communicators believed they were activing more ethically when committing an act of omission i.e. failing to disclose rather than telling a prosocial lie. Bolino and Grant (2016:599) suggest three types of prosocial decision making and behaviour:

- *prosocial motive* namely the desire to benefit others or expend effort out of concern for others;
- *prosocial behaviour* acting to promote or protect the welfare of individuals, groups, or organizations; and
- *prosocial impact* the experience of making a positive difference in the lives of others through one's work.

Therefore, the use of prosocial lying by corporate bodies either as a collective strategy or by individuals who work for such corporates is complex. Prosocial lying could thus be justified by the individual through a utilitarian perspective, i.e. their action or decision creates the greatest benefit for the most; or secondly a prescriptive, paternalistic perspective that the individual can take the action because they have more knowledge than others and can act in their best interests. Thirdly an egocentric perspective could be argued i.e. that in order to protect the organization, or the staff who work in that organization, prosocial lying is acceptable, for example, if disclosure of information could lead to loss of customer contracts, factory closures or job losses.

In the literature, prosocial behaviour in terms of corporate social responsibility (CSR) is often seen as positive (Murray and Vogel 1997; Sen and Bhattacharya 2001), however here it is postulated that prosocial behaviour is complex and what benefits one stakeholder may be detrimental to another leading to a heuristics based trade-off that can engender deception. The decision whether to engage in dishonesty or not is mediated by the trade-off between the potential benefit(s) of being dishonest and the potential personal or organizational cost of doing so (Pittarello et al. 2015) who argue that factors of influence include personal perceptions of dishonestly, and willingness to "stretch the truth".

Delmas and Burbano (2011) assert that increasing numbers of organizations, whether intentionally or unintentionally, are being misleading by engaging in greenwashing through their publications and discourse about environmental performance or the environmental benefits of a product or service. Greenwashing has been variously described as:

- The dissemination of false or incomplete information by an organization to present an environmentally responsible public image (Furlow 2010).
- The selective disclosure of positive information about an organization's environmental or social performance, without full disclosure of negative information on these dimensions, so as to create an overly positive corporate image (Lyon and Maxwell 2011),
- A strategic action in that it may mislead stakeholders about an organization's actual social performance (Roulet and Touboul 2015); or
- The tendency of a company to beautify its image through communication that emphasizes positive achievements and conceals negative conduct often to drive transient image improvement (Baldassarre and Campo 2016).

However, verified labelling, voluntary guidelines and certification programs sometimes lack substance in terms of showing anything meaningful, quantifiable or demonstrable so in themselves are a form of greenwashing (Wiseman 2018). Thus, CSR claims made by organizations can be underpinned by either substantive or symbolic aspects of corporate performance (Perez-Batres et al. 2012) and these can vary in the degree of truthfulness, trustworthiness or their veracity. Factors that mediate the process of disclosure and positive social behaviour include power dynamics, stakeholder pressure, degree of uncertainty and information asymmetry and the level of strategic greenwashing so in some instances initial symbolic claims can lead to substantive claims later in the process of corporate evolution towards sustainable practice (Berchicci and King 2007; Perez-Batres et al. 2012).

Paltering is described as "the active use of truthful statements to convey a misleading impression … [and] differs from passive lying by omission or active lying by commission i.e. the active use of false statements" (Rogers et al. 2017:456). Paltering includes activities more commonly known as fudging, twisting, shading, bending, stretching, slanting, exaggerating, distorting, whitewashing, and selective reporting (Heckman et al. 2015). Paltering is otherwise described as talking insincerely and taking advantage of customers who could be organizations or the final consumers (Stevens 1999). Is paltering an example of moral hazard in the food supply chain? Moral hazard is the risk that in a transaction, as a result of providing partial or misleading information or taking excessive risks, one party is not acting in good faith because they know that such risks are covered by insurance (Manning 2018a). Thus paltering may occur if opportunistic organizations feel they can breach agreements, especially if it is without fear of penalty (Wang et al. 2017). Does the food industry undertake paltering activity, in the form of greenwashing or selective disclosure? Is information asymmetry a driver for such behaviour? Does this context drive a demand for truth-telling and truth seeking?

9.7 Truth-Telling and Truth-Seeking

Truth-telling is simply the avoidance of dishonesty or lying. Truth-telling centers on the ethics of consumers rights to know about how their food is produced and to be able to consider the negative externalities of their purchasing decision as well as the positive benefits to themselves, the environment and wider society. Truth-telling is a concept that has been considered in detail in the medical literature (see Sullivan et al. 2001; Madhiwalla 2013) but there is scant consideration in the food and business literature thus the reason for the emphasis here. Truth-telling considers the autonomy of the consumer, i.e. it questions whether the consumer has enough information to make a reasoned and informed decision.

Truth-seeking is another term little used in the food literature. Truth-seeking is an activity that gathers, analyses and interprets information (Hair et al. 2007). Truth-seeking can also be described as the activities undertaken by consumers or those in the supply chain to identify the truth about the food they eat, the way it has been produced, manufactured and sold and the impact on individuals and society more generally. However, all market actors such as buyers, customers, consumers all make decisions under conditions of imperfect information and there is always a risk of moral hazard (Starbird 2005) so individuals may use heuristics to allow them to make decisions in what are often quite complex decisions. Heuretics are an intuitive approach that individuals or groups use when they have either quite limited information, or limited understanding of the information provided (Kahneman and Tversky 1979). Heuristics give a clear rationale to individuals on how they should make decisions or address complex problems. Thus, heuristics are "decision rules, cognitive mechanisms, and subjective opinions people use to assist them in making decisions" (Busenitz and Barney 1997:12) as they reduce complex mental tasks to simpler processes (Slovic et al. 1982). Consumers may use certain product or packaging cues as a proxy or heuristic to make otherwise complex decision making difficult especially if they are seeking to understand multi-layered issues such as sustainability and prosocial behaviour.

Sustainability labels allow consumers to use visual cues that relate to environmental and ethical considerations, and more indirect inferences can be drawn from other product characteristics identified on labels such as country of origin, or provenance (Grunert et al. 2014; Rees et al. in press). Demonstration of such attributes or characteristics on labels implies the supply chain is being truthful, honest and transparent. Corporate discourse including food labelling can be a vehicle for information disclosure in a positively prosocial endeavour or alternatively be an instrument of deception where brand equity is vulnerable to misrepresentation and mislabelling (Charlebois et al. 2016). Choice architecture is the use of informational or physical structures within the environment that influence, sometimes automatically, the way in which choices are made in order to facilitate socially desirable decisions (Thaler and Sunstein 2008; Lehner et al. 2016). Nudges are:

> interventions that aim at altering people's behaviour by either harnessing their cognitive biases or responding to them, while keeping option sets and monetary incentive structures largely intact. (Schubert 2017:330)

Nudges can be paternalistic or non- paternalistic (Schubert 2017), should be transparent (Thaler and Sunstein 2008) and not compromise an individual's autonomy or freedom of choice (Schubert 2015). However, nudges can be subtle and covert and lack transparency (Bruns et al. 2018) and include targeted information and graphics at the point of purchase, social advertising or prompted choices through choice architecture however they may use only partial or targeted disclosure (Schubert 2017). Nudges can promote prosocial goals, be educative and ensure agency (Sunstein 2015) so with regard to the environmental and social aspects of how food is produced, consumers can act as *agents for transition* especially if they are influenced by green marketing initiatives (Jaca et al. 2018). Green marketing and green nudges are aimed at encouraging pro-environmental behaviour (Schubert 2017). Chen (2010) found that there was a positive relationship between green brand equity and green brand image, and green trust. However can sustainability attributes be reduced to a single cue? Yates-Doerr (2012) terms this approach as *"the opacity of reduction*" i.e. the pretence of simplicity in terms of decision-making when actually the issue is far more complex and how this mechanism ultimately leads to greater consumer confusion. Indeed, the binary nature of explaining the characteristics of sustainability such as: good versus bad, positive versus negative can allow a multiplicity of consumer perceptions and this influences their decisions (Rees et al. in press) and how they may try to further seek the truth. The ultimate aim of transition management is to influence and create agency so that individuals can shape sustainability and as a result contribute to the needed transitions in food production and supply (Rauschmayer et al. 2015). So what value is there for corporations for engaging in CSR?

9.8   The Interaction Between Corporate Responsibility and Brand Value

There is an argument that private supply chain standards and protocols have been developed with as much, if not more emphasis on ensuring brand protection than meeting consumer requirements. For brand owners, their brand is an intangible asset (Macrae and Uncles 1997), a vehicle to create an emotional bond between consumers and their products (Fan 2005) and a tool for differentiation. Socio-technical transition results from the interaction between the exogenous landscape, the prevailing regime, and the niche (Geels 2002). Regime change is a function of selection pressure, adaptive capacity and resource allocation and the circular impact of regime change on the original selection pressures (Smith et al. 2005). There are increasing selection pressures on brand owners in the food supply chain (see Manning 2007). Emerging selection pressures are myriad but include: hybridization of public and private food governance (Verbruggen and Havinga 2015); government "Naming and Shaming" Programmes such as the Campylobacter Retail Survey Programme (FSA 2015); and increasing media interest with undercover investigations by journalists e.g. the investigation of 2 Sisters Food Group by ITN and Guardian journalists (Guardian 2018). Other selective pressures on brand value include consumer perceptions of corporate behaviour, retention of consumer trust, costs, shareholder value, the framing of consumer choice, information asymmetry, truth-seeking and meeting new trends in the supply chain such as diet transition).

Brand equity is the balance between brand assets and brand liabilities (Aaker 1991), however actors may have alternative views on what constitutes a brand asset or brand liability (Manning 2015). Brand value is co-created by different stakeholders as a result of contention and negotiation in a social process of information sharing, alignment, involvement, influence and control (Iglesias et al. 2013). Thus, CSR and corporate reputation can have a positive influence on brand equity and credibility (Lai et al. 2010; Hur et al. 2014). CSR affects brand equity, but the influence is nuanced so that individually CSR towards communities (credibility) or towards consumers (visibility) does not have a large effect on brand equity, but when combined has more influence than CSR towards other stakeholders (Torres et al. 2012). Thus whilst internal stakeholders (employees, suppliers, contractors) are important, internal negative behaviour within existing regimes may not be disclosed to other stakeholders. This failure to disclose can be a threat to brand equity. CSR projects create costs not all such projects create value especially for shareholders (Husted and Allen 2007) and this is can be a barrier to regime change. What about the role of ecopreneurship in delivering regime change and transition.

9.9   Ecopreneurship in a Transitioning Environment

Sustainable entrepreneurship or ecopreneurship is a developing field of literature (Hockerts and Wüstenhagen 2010; Galkina and Hultman 2016). Sustainable entrepreneurship is: *"the discovery and exploitation of economic opportunities through the generation of market disequilibria that initiate the transformation of a sector towards an environmentally and socially more sustainable state"* (Hockerts and Wüstenhagen 2010:482). Thus sustainable entrepreneurship goes beyond regime based CSR and provides an opportunity to transition an incumbent regime through destabilizing practices and processes to then take advantages of opportunities in the niche. Prevailing regimes face lock-in and path dependency (Geels 2004) and drive incremental innovation rather than system level step change. Indeed, prevailing regimes focus on sustainability discourse and metrics and require niche ecopreneurs to drive step changes in transition (Hockerts and Wüstenhagen 2010). Transition *through endogenous* renewal is a form of regime focused, top down transition often seeking to use the same internal resources, protocols and processes that have gone before. This often limits the success of such actions. Prevailing regimes may also re-orientate their trajectories, following a shock either internal or external to the regime (Geels and Schot 2007) in an internalised uncoordinated way. This approach to transition is often weakened by the lack of new ideas and skills that are brought in to address the problem.

   Ecopreneurs reject standard production methods, products, market structures and consumption patterns and instead develop superior sustainable products and services (Schaltegger 2002). Thus, ecopreneurs are agents of change in the transitioning socio-technical environment. An *emergent transformation* sees ecopreneurs in a "bottom up" uncoordinated response seek to drive transition either in isolation or in consort with the prevailing regime. These latter business environments are increasingly becoming new socio-technical networks of practice (Letaifa 2014) based on integrity and associated social value creation. *Purposive transitions* are an intended symbiotic response between the prevailing regime and ecopreneurial niches to co-evolve and co-create new production methods, products and market structures. As a result, purposive transition networks are based on relational, interaction based general trust. A food system is: *"a sociotechnical regime made up of dominant economic, industrial, political and scientific rules and assumptions"* (Marsden 2013:124). These rules and assumptions connect with truth-values, moral hazard and potential for one actor misleading another. There are two kinds of endogenous processes that lead to rule change (Geels and Schot 2007). The first, they argue is evolutionary-economic were the rules change indirectly through market driven product variation, for example nutrition transition and diet transition to address GHGEs. The second is social-institutional where actors directly negotiate the rules in networks of practice. This latter process is driven by discourse and engagement with groups interacting, sensemaking, building belief systems and reaching closure and consensus of a shared cognitive frame (Geels and Schot 2007). A process where tuth0telling and truth-seeking is key.

9.10 Conclusion

Transitioning from the prevailing food regime to another more sustainable socio-technical system will be driven by multiple factors primarily consumers seeking truth and transparency with the foods they purchase and consume and ecopreneurial niches affecting prevailing regime path dependency and an unwillingness to change. This evolving socio-technical symbiosis will be required to meet the environmental pressures and socio-economic driver that influence the food supply chain and ensure that the growing human population can be nourished into the future.

References

Aaker, D. A. (1991). *Managing brand equity; capitalizing on the value of a brand name*. New York: The Free Press.

Alexander, P., Brown, C., Arneth, A., Finn igan, J., & Rounsevell, M. D. (2016). Human appropriation of land for food: The role of diet. *Global Environmental Change, 41*, 88–98.

Ali, M. H., & Suleiman, N. (2018). Eleven shades of food integrity: A halal supply chain perspective. *Trends in Food Science and Technology, 71*, 216–224.

Andrikopoulos, A. (2015). Truth and financial economics: A review and assessment. *International Review of Financial Analysis, 39*, 186–195.

Avelino, F., & Rotmans, J. (2009). Power in transition: An interdisciplinary framework to study power in relation to structural change. *European Journal of Social Theory, 12*(4), 543–569.

Badham, R., Clegg, C., & Wall, T. (2001). Socio-technical theory. In W. Karwowski (Ed.), *International encyclopedia of ergonomics and human factors* (pp. 1370–1373). London: Taylor & Francis.

Baldassarre, F., & Campo, R. (2016). Sustainability as a marketing tool: To be or to appear to be? *Business Horizons, 59*, 421–429.

Berchicci, L., & King, A. (2007). 11 postcards from the edge: A review of the business and environment literature. *The Academy of Management Annals, 1*(1), 513–547.

Berger, P. L., & Luckmann, T. (1991). *The social construction of reality: A treatise in the sociology of knowledge*. (No. 10). UK: Penguin.

Berkhout, F., Smith, A., & Stirling, A. (2004). Socio-technological regimes and transition contexts. In B. Elzen, F. W. Geels, & K. Green (Eds.), *System innovation and the transition to sustainability: Theory, evidence and policy* (pp. 48–75). Cheltenham: Edward Elgar.

Blumer, H. (1986). *Symbolic interactionism: Perspective and method.* Berkeley: University of California Press.

Bolino, M. C., & Grant, A. M. (2016). The bright side of being prosocial at work, and the dark side, too: A review and agenda for research on other-oriented motives, behavior, and impact in organizations. *The Academy of Management Annals, 10*(1), 599–670.

Bruns, H., Kantorowicz-Reznichenko, E., Klement, K., Jonsson, M. L., & Rahali, B. (2018). Can nudges be transparent and yet effective? *Journal of Economic Psychology, 65*, 41–59.

Burggraf, C., Kuhn, L., Zhao, Q. R., Teuber, R., & Glauben, T. (2015). Economic growth and nutrition transition: An empirical analysis comparing demand elasticities for foods in China and Russia. *Journal of Integrative Agriculture, 14*(6), 1008–1022.

Busenitz, L. W., & Barney, J. B. (1997). Differences between entrepreneurs and managers in large organizations: Biases and heuristics in strategic decision-making. *Journal of Business Venturing, 12*(1), 9–30.

Cao, C., Xia, C., & Chan, K. C. (2016). Social trust and stock price crash risk: Evidence from China. *International Review of Economics and Finance, 46*, 148–165.

Chance, D., Cicon, J., & Ferris, S. P. (2015). Poor performance and the value of corporate honesty. *Journal of Corporate Finance, 33*, 1–18.

Charlebois, S., Schwab, A., Henn, R., & Huck, C. W. (2016). Food fraud: An exploratory study for measuring consumer perception towards mislabeled food products and influence on self-authentication intentions. *Trends in Food Science & Technology, 50*, 211–218.

Chen, Y. S. (2010). The drivers of green brand equity: Green brand image, green satisfaction, and green trust. *Journal of Business Ethics, 93*(2), 307–319.

Choi, H. J., Park, H. S., & Oh, J. Y. (2011). Cultural differences in how individuals explain their lying and truth-telling tendencies. *International Journal of Intercultural Relations, 35*(6), 749–766.

Cohen, N., & Ilieva, R. T. (2015). Transitioning the food system: A strategic practice management approach for cities. *Environmental Innovation and Societal Transitions, 17*, 199–217.

Dagevos, H., & Voordouw, J. (2013). Sustainability and meat consumption: Is reduction realistic? *Sustainability: Science, Practice and Policy, 9*(2), 60–69.

Davidson, D. (2001). *Truth and meaning. Inquiries into truth and interpretation* (pp. 17–36). Oxford: Oxford University Press.

Deaver, F. (1990). On defining truth. *Journal of Mass Media Ethics, 5*, 168–177.

Delmas, M. A., & Burbano, V. C. (2011). The drivers of greenwashing. *California Management Review, 54*(1), 64–87.

Devin, B. (2016). Half-truths and dirty secrets: Omissions in CSR communication. *Public Relations Review, 42*(1), 226–228.

Dobler, M. (2008). Incentives for risk reporting – A discretionary disclosure and cheap talk approach. *The International Journal of Accounting, 43*(1), 184–206.

Dwivedi, S. L., van Bueren, E. T. L., Ceccarelli, S., Grando, S., Upadhyaya, H. D., & Ortiz, R. (2017). Diversifying food systems in the pursuit of sustainable food production and healthy diets. *Trends in Plant Science, 22*(10), 842–856.

Emery, F. Y. T., & Trist, E. (1960). *Socio-technical systems. Management science models and techniques.* London: Pergamon Press.

Emery, F. E., & Trist, E. L. (1965). The causal texture of organizational environments. *Human Relations, 18*(1), 21–32.

Fan, Y. (2005). Ethical branding and corporate reputation. *Corporate Communications: An International Journal, 10*(4), 341–350.

Fisher, R. (2013). 'A gentleman's handshake': The role of social capital and trust in transforming information into usable knowledge. *Journal of Rural Studies, 31*, 13–22.

Friedmann, H. (1993). The political economy of food: A global crisis. *New Left Review, 197*, 29–57.

FSA (Food Standards Agency). (2015). Year 1 of a UK-wide survey of Campylobacter contamination on fresh chickens at retail (February 2014 to February 2015). Resource document. https://

www.food.gov.uk/other/year-1-of-a-uk-wide-survey-of-campylobacter-contamination-on-fresh-chickens-at-retail-february-2014-to-february-2015. Accessed 9 June 2018.

Furlow, N. E. (2010). Greenwashing in the new millennium. *The Journal of Applied Business and Economics, 10*(6), 22.

Galeotti, F., Kline, R., & Orsini, R. (2017). When foul play seems fair: Exploring the link between just deserts and honesty. *Journal of Economic Behavior & Organization, 142*, 451–467.

Galkina, T., & Hultman, M. (2016). Ecopreneurship–assessing the field and outlining the research potential. *Small Enterprise Research, 23*(1), 58–72.

Geels, F. W. (2002). Technological transitions as evolutionary reconfiguration processes: A multi-level perspective and a case-study. *Research Policy, 31*(8–9), 1257–1274.

Geels, F. W. (2004). From sectoral systems of innovation to socio-technical systems: Insights about dynamics and change from sociology and institutional theory. *Research Policy, 33*(6–7), 897–920.

Geels, F. W. (2010). Ontologies, socio-technical transitions (to sustainability), and the multi-level perspective. *Research Policy, 39*, 495–510.

Geels, F. W., & Schot, J. (2007). Typology of sociotechnical transition pathways. *Research Policy, 36*(3), 399–417.

Geels, F. W., Sovacool, B. K., Schwanen, T., & Sorrell, S. (2017). The socio-technical dynamics of low-carbon transitions. *Joule, 1*(3), 463–479.

Ghaffarian, V. (2011). The new stream of socio-technical approach and main stream information systems research. *Procedia Computer Science, 3*, 1499–1511.

Glanzberg, M. (2014). Truth. In E. Z. Fall (Ed.), *Stanford encyclopedia of philosophy*. Stanford: Stanford University.

Gneezy, U. (2005). Deception: The role of consequences. *American Economic Review, 95*(1), 384–394.

Goddard, E., Muringai, V., & Boaitey, A. (2018). Food integrity and food technology concerns in Canada: Evidence from two public surveys. *Journal of Food Quality, 2018*, 1–12.

Green, S. P., & Kugler, M. B. (2012). Public perceptions of white collar crime culpability: Bribery, perjury, and fraud. *Law and Contemporary Problems, 75*(2), 33–59.

Grin, J., Rotmans, J., & Schot, J. (2010). *Transitions to sustainable development: New directions in the study of long term transformative change*. New York: Routledge.

Grover, S. L. (1993). Lying, deceit, and subterfuge: A model of dishonesty in the workplace. *Organization Science, 4*(3), 478–495.

Grunert, K. G., Hieke, S., & Wills, J. (2014). Sustainability labels on food products: Consumer motivation, understanding and use. *Food Policy, 44*, 177–189.

Hair, J. F., Money, A. H., Samouel, P., & Page, M. (2007). *Research methods for business*. Chichester: Wiley.

Hartmann, M., Klink, J., & Simons, J. (2015). Cause related marketing in the German retail sector: Exploring the role of consumers' trust. *Food Policy, 52*, 108–114.

Heckman, K. E., Stech, F. J., Thomas, R. K., Schmoker, B., & Tsow, A. W. (2015). *Cyber denial, deception and counter deception*. Basel: Springer.

Hockerts, K., & Wüstenhagen, R. (2010). Greening Goliaths versus emerging Davids—theorizing about the role of incumbents and new entrants in sustainable entrepreneurship. *Journal of Business Venturing, 25*(5), 481–492.

Hur, W. M., Kim, H., & Woo, J. (2014). How CSR leads to corporate brand equity: Mediating mechanisms of corporate brand credibility and reputation. *Journal of Business Ethics, 125*(1), 75–86.

Husted, B. W., & Allen, D. B. (2007). Strategic corporate social responsibility and value creation among large firms: Lessons from the Spanish experience. *Long Range Planning, 40*(6), 594–610.

Iglesias, O., Ind, N., & Alfaro, M. (2013). The organic view of the brand: A brand value co-creation model. *Journal of Brand Management, 20*(8), 670–688.

Jaca, C., Prieto-Sandoval, V., Psomas, E. L., & Ormazabal, M. (2018). What should consumer organizations do to drive environmental sustainability? *Journal of Cleaner Production, 181*, 201–208.

Jarosz, L. (2008). The city in the country: Growing alternative food networks in Metropolitan areas. *Journal of Rural Studies, 24*(3), 231–244.

Johnstone, P., & Newell, P. (2018). Sustainability transitions and the state. *Environmental Innovation and Societal Transitions, 27*, 72–82.

Kahneman, D., & Tversky, A. (1979). Prospect theory: An analysis of decision under risk. *Econometrica, 47*, 263–291.

Kasperson, R. E., & Golding, D. S. T. (1992). Social distrust as a factor in siting hazardous facilities and communicating risks. *Journal of Social Issues, 48*, 161–187.

Krzyzaniak, S. A. C. (2018). *Determining the barriers to effective food safety governance in food manufacturing: A case study.* (Doctoral dissertation, University of Portsmouth). Available at: https://ethos.bl.uk/OrderDetails.do?uin=uk.bl.ethos.765711.

Kuokkanen, A., Nurmi, A., Mikkilä, M., Kuisma, M., Kahiluoto, H., & Linnanen, L. (2018). Agency in regime destabilization through the selection environment: The Finnish food system's sustainability transition. *Research Policy, 47*(8), 1513–1522.

Laakso, S., & Lettenmeier, M. (2016). Household-level transition methodology towards sustainable material footprints. *Journal of Cleaner Production, 132*, 184–191.

Lai, C. S., Chiu, C. J., Yang, C. F., & Pai, D. C. (2010). The effects of corporate social responsibility on brand performance: The mediating effect of industrial brand equity and corporate reputation. *Journal of Business Ethics, 95*(3), 457–469.

Lawhon, M., & Murphy, J. T. (2012). Socio-technical regimes and sustainability transitions: Insights from political ecology. *Progress in Human Geography, 36*(3), 354–378.

Lehner, M., Mont, O., & Heiskanen, E. (2016). Nudging–a promising tool for sustainable consumption behaviour? *Journal of Cleaner Production, 134*, 166–177.

Letaifa, S. B. (2014). The uneasy transition from supply chains to ecosystems: The value-creation/value-capture dilemma. *Management Decision, 52*(2), 278–295.

Levidow, L. (2015). European transitions towards a corporate-environmental food regime: Agroecological incorporation or contestation? *Journal of Rural Studies, 40*, 76–89.

Levine, E., Hart, J., Moore, K., Rubin, E., Yadav, K., & Halpern, S. (2018). The surprising costs of silence: Asymmetric preferences for prosocial lies of commission and omission. *Journal of Personality and Social Psychology, 114*(1), 29.

Liao, C., & Brown, D. G. (2018). Assessments of synergistic outcomes from sustainable intensification of agriculture need to include smallholder livelihoods with food production and ecosystem services. *Current Opinion in Environmental Sustainability, 32*, 53–59.

Lupoli, M. J., Levine, E. E., & Greenberg, A. E. (2018). Paternalistic lies. *Organizational Behaviour and Human Decision Processes, 148*, 31–50.

Lyon, T. P., & Maxwell, J. W. (2011). Greenwash: Corporate environmental disclosure under threat of audit. *Journal of Economics and Management Strategy, 20*(1), 3–41.

Macrae, C., & Uncles, M. D. (1997). Rethinking brand management: The role of brand chartering. *Journal of Product and Brand Management, 6*(1), 64–77.

Madhiwalla, N. (2013). The ethics of truth telling. *South Asian Journal of Cancer, 2*(2), 53.

Manning, L. (2007). Food safety and brand equity. *British Food Journal, 109*(7), 496–510.

Manning, L. (2015). Determining value in the food supply chain. *British Food Journal, 117*(11), 2649–2663.

Manning, L. (2018a). Food supply chain fraud: The economic environmental and socio-political consequences. In D. Barling & J. Fanzo (Eds.), *Advances in food security and sustainability* (Vol. 3, pp. 253–276). London: Academic.

Manning, L. (2018b). Triangulation: Effective verification of food safety and quality management systems and associated organizational culture. *Worldwide Hospitality and Tourism Themes, 10*(3), 297–312.

Markard, J., Raven, R., & Truffer, B. (2012). Sustainability transitions: An emerging field of research and its prospects. *Research Policy, 41*(6), 955–967.

Marsden, T. (2013). From post-productionism to reflexive governance: Contested transitions in securing more sustainable food futures. *Journal of Rural Studies, 29*, 123–134.

Maye, D., & Kirwan, J. (2010). Alternative food networks. *Sociology of Agriculture and Food, 20*, 383–389.

Meynard, J. M., Jeuffroy, M. H., Le Bail, M., Lefèvre, A., Magrini, M. B., & Michon, C. (2017). Designing coupled innovations for the sustainability transition of agrifood systems. *Agricultural Systems, 157*, 330–339.

Mitchell, F., Nørreklit, H., & Nørreklit, L. (2017). The validity of financial statement measurement. In H. Nørreklit (Ed.), *A philosophy of management accounting* (pp. 134–148). New York: Routledge.

Mumford, E. (2006). The story of socio-technical design: Reflections on its successes, failures and potential. *Information Systems Journal, 16*(4), 317–342.

Murray, K. B., & Vogel, C. M. (1997). Using a hierarchy of effects approach to gauge the effectiveness of CSR to generate goodwill towards the firm: Financial versus nonfinancial impacts. *Journal of Business Research, 38*(2), 141–159.

Nørreklit, L., Jack, L., & Nørreklit, H. (2018). Beyond the post-truth turn: From habitus based to paranoiac based performance management. *Proceedings of Pragmatic Constructivism, 8*(1), 17–18.

O'Connor, A., & Meister, M. (2008). Corporate social responsibility attribute rankings. *Public Relations Review, 34*(1), 49–50.

Paül, V., & McKenzie, F. H. (2013). Peri-urban farmland conservation and development of alternative food networks: Insights from a case-study area in metropolitan Barcelona (Catalonia, Spain). *Land Use Policy, 30*(1), 94–105.

Perez-Batres, L. A., Doh, J. P., Miller, V. V., & Pisani, M. J. (2012). Stakeholder pressures as determinants of CSR strategic choice: Why do firms choose symbolic versus substantive self-regulatory codes of conduct? *Journal of Business Ethics, 110*(2), 157–172.

Peters, R., Covello, V., & McCallum, D. (1997). The determinants of trust and credibility in environmental risk communication. *Risk Analysis, 17*, 43–54.

Pigors, M., & Rockenbach, B. (2016). The competitive advantage of honesty. *European Economic Review, 89*, 407–424.

Pittarello, A., Leib, M., Gordon-Hecker, T., & Shalvi, S. (2015). Justifications shape ethical blind spots. *Psychological Science, 26*(6), 794–804.

Popkin, B. M. (2001). The nutrition transition and obesity in the developing world. *The Journal of Nutrition, 131*, 871–873.

Popkin, B. M. (2002). The shift in stages of the nutrition transition in the developing world differs from past experience. *Public Health Nutrition, 5*, 205–214.

Popkin, B. M. (2007). Understanding global nutrition dynamics as a step towards controlling cancer incidence. *Nature Reviews Cancer, 7*, 61–67.

Popkin, B. M. (2008). Will China's nutrition transition overwhelm its health care system and slow economic growth? *Health Affairs, 27*, 1064–1076.

Ranganathan, J., Vennard, D., Waite, R., Dumas, P., Lipinski, B., & Searchinger, T. (2016). *Shifting diets for a sustainable food future*. Washington, DC: World Resources Institute.

Rauschmayer, F., Bauler, T., & Schäpke, N. (2015). Towards a thick understanding of sustainability transitions—linking transition management, capabilities and social practices. *Ecological Economics, 109*, 211–221.

Rees, W., Tremma, O., & Manning, L. (in press). Sustainability cues on packaging: The influence of recognition on purchasing behaviour. (Under review).

Regan, Á., Marcu, A., Shan, L. C., Wall, P., Barnett, J., & McConnon, Á. (2015). Conceptualising responsibility in the aftermath of the horsemeat adulteration incident: An online study with Irish and UK consumers. *Health, Risk & Society, 17*(2), 149–167.

Rogers, T., Zeckhauser, R., Gino, F., Norton, M. I., & Schweitzer, M. E. (2017). Artful paltering: The risks and rewards of using truthful statements to mislead others. *Journal of Personality and Social Psychology, 112*(3), 456–473.

Rosenbaum, S. M., Billinger, S., & Stieglitz, N. (2014). Let's be honest: A review of experimental evidence of honesty and truth-telling. *Journal of Economic Psychology, 45*, 181–196.

Rotmans, J., & Loorbach, D. (2009). Complexity and transition management. *Journal of Industrial Ecology, 13*(2), 184–196.

Rotmans, J., & Loorbach, D. (2010). Towards a better understanding of transitions and their governance. A systemic and reflexive approach. In J. Grin, J. Rotmans, & J. Schot (Eds.), *Transitions to sustainable development—new directions in the study of long term transformation change* (pp. 105–198). New York: Routledge.

Rotmans, J., Kemp, R., & Asselt, M. V. (2001). More evolution than revolution: Transition management in public policy. *Foresight, 3*(1), 15–31.

Roulet, T. J., & Touboul, S. (2015). The intentions with which the road is paved: Attitudes to liberalism as determinants of greenwashing. *Journal of Business Ethics, 128*(2), 305–320.

Schaltegger, S. (2002). A framework for ecopreneurship. *Greener Management International, 38*, 45–58.

Schösler, H., & de Boer, J. (2018). Towards more sustainable diets: Insights from the food philosophies of "gourmets" and their relevance for policy strategies. *Appetite, 127*, 59–68.

Schreck, P. (2015). Honesty in managerial reporting: How competition affects the benefits and costs of lying. *Critical Perspectives on Accounting, 27*, 177–188.

Schubert, C. (2015). On the ethics of public nudging: Autonomy and agency. Resource document. https://papers.ssrn.com/sol3/papers.cfm?abstract_id=2672970. Accessed 31 Jul 2018.

Schubert, C. (2017). Green nudges: Do they work? Are they ethical? *Ecological Economics, 132*, 329–342.

Sen, S., & Bhattacharya, C. B. (2001). Does doing good always lead to doing better? Consumer reactions to corporate social responsibility. *Journal of Marketing Research, 38*(2), 225–243.

Silva, A., & Stocker, L. (2018). What is a transition? Exploring visual and textual definitions among sustainability transition networks. *Global Environmental Change, 50*, 60–74.

Slovic, P., Fischhoff, B., & Lichtenstein, S. (1982). Why study risk perception? *Risk Analysis, 2*(2), 83–93.

Smink, M. M., Hekkert, M. P., & Negro, S. O. (2015). Keeping sustainable innovation on a leash? Exploring incumbents' institutional strategies. *Business Strategy and the Environment, 24*(2), 86–101.

Smith, A., Stirling, A., & Berkhout, F. (2005). The governance of sustainable socio-technical transitions. *Research Policy, 34*(10), 1491–1510.

Snellman, J. W. (1911). The meaning and test of truth. *Mind, 20*(78), 235–242.

Starbird, S. A. (2005). Moral hazard, inspection policy, and food safety. *American Journal of Agricultural Economics, 87*(1), 15–27.

Stevens, B. (1999). Persuasion, probity, and paltering: The Prudential crisis. *The Journal of Business Communication (1973), 36*(4), 319–334.

Sullivan, R. J., Menapace, L. W., & White, R. M. (2001). Truth-telling and patient diagnoses. *Journal of Medical Ethics, 27*(3), 192–197.

Sunstein, C. R. (2015). Nudges do not undermine human agency. *Journal of Consumer Policy, 38*(3), 207–210.

Thaler, R. H., & Sunstein, C. R. (2008). *Nudge: Improving decisions about health, wealth, and happiness*. New Haven: Yale University Press.

The Guardian. (2018). 2 Sisters guilty of poor hygiene at poultry plants, FSA finds. (Friday 2 March 2018) Resource document. https://www.theguardian.com/business/2018/mar/02/2-sisters-guilty-of-poor-hygiene-at-poultry-plants-fsa-finds. Accessed 9 June 2018.

Torres, A., Bijmolt, T. H., Tribó, J. A., & Verhoef, P. (2012). Generating global brand equity through corporate social responsibility to key stakeholders. *International Journal of Research in Marketing, 29*(1), 13–24.

Verbruggen, P., & Havinga, T. (2015). Food safety meta-controls in The Netherlands. *European Journal of Risk Regulation, 6*, 512–524.

Vergragt, P. J., Dendler, L., de Jong, M., & Matus, K. (2016). Transitions to sustainable consumption and production in cities. *Journal of Cleaner Production, 134*, 1–12.

Vittersø, G., & Tangeland, T. (2015). The role of consumers in transitions towards sustainable food consumption. The case of organic food in Norway. *Journal of Cleaner Production, 92*, 91–99.

Wang, C. S., Van Fleet, D. D., & Mishra, A. K. (2017). Food integrity: A market-based solution. *British Food Journal, 119*(1), 7–19.

Winter, S., Berente, N., Howison, J., & Butler, B. (2014). Beyond the organizational 'container': Conceptualizing 21st century sociotechnical work. *Information and Organization, 24*(4), 250–269.

Wiseman, S. R. (2018). Localism, labels, and animal welfare. *Northwestern Journal of Law & Social Policy, 13*(2), 66.

Yates-Doerr, E. (2012). The opacity of reduction: Nutritional black-boxing and the meanings of nourishment. *Food, Culture & Society, 15*(2), 293–313.

Conclusions

There are a lot of issues in agriculture-related fields right now, as we have high-lighted in the introduction chapter, two of which have re-emerged quite strongly from the chapters in this book: one relates to environment, to the disrupted ecological processes and our very existence on planet Earth, as we rely on resources that come from nature; the second revolves around the continuum between production and consumption, for which much of the rhetoric along the lines of 'sustainability' and 'feeding the world' is quite controversial. This is what makes it a transition issue, as we feel the pressure of responding to them and address the challenges it imposes in order to move to better provisioning systems. In this, technologies are at the forefront of the transition while constituting a concerning factor; at the beginning of the book we highlighted that current research in agri-food focuses on technology and the strands that belong to full connectivity and data ownership, and there is no doubt that the questions our colleagues ask are of course valid also in this context, as Chap. 7 has started to address. Although the focus of this book is not food security, we have to acknowledge, as Chaps. 1 and 8 did, that if there is a challenge for which food technologies are considered the pivot and even the only solution, that is about feeding the world, and do it sustainably. Truth is, responding to the call to find a master plan to feed the future nine million population is extremely complex. We need nutritious and sustainable food: can food tech deliver it? That is because some of the relationships that go unnoticed in our daily activities of production and consumption are still rooted in past relations and old structures, specifically the one pertaining to developed and developing countries (Friedmann and McMichael 1989). There are two sides of the story, and they relate to feeding the populations of both developed and developing countries addressing the specific challenges each have. Here, maybe a new trade-off of technology against resources may have a potential to respond partly to the challenge. The developed countries are struggling with lack of diversity in the food resources and rise of growing demand from the consumer side for the food products produced according to their growing ethical values, environmental and health concern (e.g. rise obesity and type II diabetics, rise of lactose intolerance, and food allergies). In these countries on the one

© Springer Nature Switzerland AG 2019
C. Piatti et al. (eds.), *Food Tech Transitions*,
https://doi.org/10.1007/978-3-030-21059-5

hand addressing the rising demands of this kind of foods depends on environmentally friendly technological solutions, while on the other hand introduction of alternative natural resources - particularly plant resources - with health benefits for food products may present significant potentials to fulfil the consumers demand. Accordingly, the main objective of Chap. 4 was to address the benefits and potentials of neotropical pseudo-cereals in terms of their nutritional and functional attributes. Chapter 5 explained the market demand for various underutilized crops as alternative and rising stars in the sky of some of the present trends in the food supply chain, namely superfoods. Such food commodities can be imported from their original places (and projects are set to introduce their cultivation massively in Europe, to take the example of soy) and in our existing food-chains they are available under the score of healthy foods, mostly in organic shops or through online markets; the re-introduction and re-discovery of these raw materials to the primary production sector of food supply chain may be among the set of tactics which can be put in place to secure a diverse and nutrition-dense future food production in a sustainable way. These are plants or, respectively, raw materials which can be regarded as a source of stable food with remarkable nutritional benefits. They are 'natural' raw plant food materials which can be used to enhance the nutritional security of people with specific dietary requirements (e.g. gluten-free, lactose free, low fat, healthy fat, low sugar, etc.) or particular diets (e.g. vegetarian, and vegan). As mentioned in Chaps. 1, 7 and 8 a personalized nutrition is now the new normal and will be stable in the future, in this plant-based food and plant-based alternatives will be foundational (although, as questioned, this latter will come to terms with concepts of 'naturalness' quite soon). But it is better bearing in mind that although such raw materials have been part of the diets of native populations in their original countries and for long time, there are multidimensional factors at play for their adaptation and inclusion in the global food production and system and consequently the global diet. To properly introduce them into the diet of populations beyond their origins, more studies about breeding and development of suitable agricultural management systems as well as developing suitable genotypes for the new environmental conditions must be carried out. Secondly, modification of current technologies in food industry and processing sector and food product development based on the new row material should be a primary target, but conditions of production, as analysed in Chap. 6, constitutes a steady barrier, both in material and physical terms, and in those relations of production that agitates economic and political debates. Although local production of such grains would remarkably shorten the food supply chain, which may be a more sustainable way of production including less transport and thus may be trustworthy for consumers willing to buy an environmentally sustainable product, some issues rooted in historical relationships should constitute a reason for concern. In this respect, developing regions such as South America, Asia and Africa are constantly asked to assist the developed countries by sharing their underutilized diverse food plant crops and rich genetic sources, often using market justifications for that. Many producers and developers in good faith think that natural resources that had been forgotten for a long time can be used for rural development in developing countries and help solving the health and environmental emergency of developed countries, pushing for more technological innovation and, on top of that, proceeding with developing novel resources such as use of genetically modified crops. In fact, at the other side of the coin, some regions of the world are in urgent need to develop more sustainable production methods both in primary production and secondary production. This of course might not be the right, ethical, sustainable thing to do, despite the good intentions (as known, the introduction of, let's say, quinoa, avocado or chia in Western markets has meant displacement of rural populations and a heavy impact on local environments and diets in developed countries). In this regard, breeding programs and the development of grains and crops with high nutritional quality as well as being adoptable to climate change is essential for these regions. As addressed in Chap. 3, reduction of agricultural losses in these regions is a primary concern, as it contributes to more efficient use of resources. In fact, post-harvest technologies discussed in this chapter can be a potential solution. On the secondary productions, as discussed in Chap. 2, novel thermal and non-thermal technologies which are more efficient and sustainable in terms of energy and water consumption may be used for further processing of agricultural production to confront the population with a more diverse food product range while extending the shelf life of the food material and increase their safety to a great extent. Novel technologies adopted to use solar or wind energy for refrigeration, heating, drying, and other forms of food processing may be trail-made for such regions. Optimistically we might argue that we have what is required to at least partially secure the food for the future populations by using only our natural resources, and research contributing to development of novel resources may be complementary to it. For making this possible however, policies in play might be among the most important factors. As it was discussed in Chap. 5, it should be noted how the rising trends dictated to a great extent by consumers and food industry must address the demand if they want to survive. Are rising trends planned to fulfil and stabilize the economical demands of market for a temporary time? The food industry had already lost the trust of a huge share of consumers as result of failures in the past, as Chap. 1 has proposed. Hence, the main players in the food supply chain should be careful to re-stabilize its operations and the consumer trust. Whether consumers demand and choice for such trends are affected by marketing is definitely the case, as Chap. 9 confirms, and for which the industry, with the massive role of corporations, has a lot to work for. Transparency is a must for survival in the current and future food supply chain. The emergence and growing stability of so called alternative food networks (Goodman and Goodman, 2009), highlighted in Chaps. 1 and 9, is evident to consumers' demand for shorter, more fair, more environmental friendly, more nutritious, and more cultural-cautious food supply chains. Corporations might easily rebrand themselves following the trends survey companies sell them. Nestlé for instance is now 'the world's leading nutrition, health and wellness company', as they proudly announce in their website, and is quite active in spreading future food scenarios, which they want of course to control. The Swiss company is also an apt example for their politics in relation to data, as they have heavily invested on e-commerce and are able to collect data which will be used extensively to reinforce their market positions. Modern technology is already playing a huge role along the supply chain, both in production as the chapters in the first part of the book have documented, and to the consumer end, as emerges from chapters in the second part. Cohen (2000:204) made a distinction between citizen-consumers and customer-consumers, proposing that the former are "consumers who take on the political responsibility we usually associate with citizens to consider the general good of the nation through their consumption, and the latter being consumers who seek primarily to maximize their personal economic interests in the market place". In the context of data collection this is not the case; experts (Selby 2017; Morozov 2018) maintain that this is an issue, as citizens, understood only as consumers, have no control over their data. The example of Barcelona, Spain, can be useful: the council has implemented a project, called DECODE which is a pilot involving three other European cities, and in which data publicly collected are going to be retained by the Council and then used in a public and transparent way to make profits to be then reinvested in the city (DECODE, 2018). As stated in their website, the control loss that many citizens feel over the personal information on the internet and as collected by the many connected technologies that rely on it is a concern, as companies profit even more from that. The project "will explore how to build a data-centric digital economy where data that is generated and gathered by citizens, the Internet of Things (IoT), and sensor networks is available for broader communal use, with appropriate privacy protections. As a result, innovators, startups, NGOs, cooperatives, and local communities can take advantage of that data to build apps and services that respond to their needs and those of the wider community" (DECODE, 2018). Can we imagine anything like this to be developed specifically in the agricultural field? We have indeed the examples and the means to do it. What we ask to technologies, how can we make technologies at the service of citizens instead of just being a tool in the hands of few, or how any data related to food transformation, processing and consumption to be used for the good of field-producers and consumers, is not be envisaged yet but will become a growing concern for the years to come.

References

Cohen, L. (2000). Citizens and consumers in the United States in the century of mass consumption. In M. Daunton & M. Hilton (Eds.), *The politics of consumption: Material culture and citizenship in Europe and America* (pp. 203–222). Oxford: Berg.
DECODE. (2018). Resource document. https://www.decodeproject.eu/blog/connecting-decode-barcelona-city-council-data-infrastructure. Accessed 18 January 2019.
Friedmann, H., & McMichael, P. (1989). Agriculture and the state system: The rise and decline of national agricultures, 1870 to the present. *Sociologia Ruralis, 29*(2), 93–117.
Goodman, D., & Goodman, M. (2009). Alternative food networks. In R. Kitchin & N. Thrift (Eds.), *International encyclopedia of human geography 3* (pp. 208–220). London: Elsevier.
Morozov, E. (2018). Reasserting cyber sovereignty: how states are taking back control. Resource document. https://www.theguardian.com/technology/2018/oct/07/states-take-back-cyber-control-technological-sovereignty. Accessed 10 December 2018.
Selby, J. (2017). Data localization laws: Trade barriers or legitimate responses to cybersecurity risks, or both? *International Journal of Law and Information Technology, 25*, 213–232.

Printed in the United States
By Bookmasters